Franziska von Verschuer
Collecting Seeds, Assembling Worlds

Science Studies

Franziska von Verschuer is a sociologist in the Biotechnology, Nature and Society research group at Goethe University Frankfurt, where she received her PhD in 2024. She studied sociology and psychology in Freiburg, Frankfurt, and Berlin. Her research and teaching focus on environmental and post-anthropocentric sociology, science and technology studies, as well as feminist and decolonial theory and epistemology.

Franziska von Verschuer

Collecting Seeds, Assembling Worlds

An Inquiry into Agrobiodiversity Conservation Through the Svalbard Global Seed Vault

[transcript]

D.30

The open access publication of this book was funded by the Open Access Publication Fund of Goethe University Frankfurt am Main.

Bibliographic information published by the Deutsche Nationalbibliothek
The Deutsche Nationalbibliothek lists this publication in the Deutsche Nationalbibliografie; detailed bibliographic data are available online at https://dnb.dnb.de

transcript Verlag | Hermannstraße 26 | D-33602 Bielefeld | live@transcript-verlag.de

Cover design: Maria Arndt
Cover photo: Mari Tefre / Tefreproduksjoner
Printing: Elanders Waiblingen GmbH, Waiblingen
https://doi.org/10.14361/9783839412596
Print-ISBN: 978-3-8376-7811-6 | PDF-ISBN: 978-3-8394-1259-6
ISSN of series: 2703-1543 | eISSN of series: 2703-1551

Printed on permanent acid-free text paper.

For Katja

Contents

Introduction

> While many may ponder the consequences of global warming, perhaps the biggest single environmental catastrophe in human history is unfolding in the garden. […] Loss of genetic diversity in agriculture – silent, rapid, inexorable – is leading us to a rendezvous with extinction – to the doorstep of hunger on a scale we refuse to imagine.
> (Fowler/Mooney 1990: ix)

> The Svalbard Global Seed Vault is a key component of global efforts to secure the diversity of our most important crop plants to ensure that they meet our needs for the long term […]. It is a more complex and nuanced story than the "Doomsday Vault" romanticized by the popular press, but in the end the Svalbard Global Seed Vault is an elegantly simple and pragmatic response to a pressing global need.
> (Sir Peter Crane FRS in Fowler 2016: 14)

In September 2015, a delegation of Syrian scientists from the International Center for Agricultural Research in the Dry Areas (ICARDA) travelled to the Arctic island of Spitsbergen in the Svalbard archipelago to retrieve boxes of seeds they had deposited in the Svalbard Global Seed Vault over the course of the previous seven years. A series of droughts had challenged agricultural production in Syria between 2006 and 2009 and given rise to a food crisis exacerbating existing political, economic, and social unrest. All this triggered mass protests against the authoritarian national government in 2011 that developed into a civil war within little more than a year. To this day, this war has caused hundreds of thousands of deaths and millions of refugees, and devastated societal infrastructure, cultural heritage, and ecological habitats. With the loss of Syria's national seed genebank in Tel Hadia, a few kilometres south of Aleppo, it has also claimed a globally unique natural heritage: "the largest collection of crop diversity from the earliest centre of origin of agriculture, the Fertile Crescent [...] important not only for agriculture in the Middle East but for global food production and food security" (Westengen et al. 2020: 1311). Before the genebank staff were forced to abandon the facility in 2014, however, ICARDA had managed to send duplicates of 83 per cent of this collection to the Svalbard Global Seed Vault, a seed conservation facility in the Arctic that offers secure storage of safety duplicates of seed collections from all over the world. Beginning in 2015, when remotely monitoring the state of the occupied genebank in Tel Hadia was no longer possible and recovering the original seed collection stored there no longer seemed a realistic prospect, ICARDA began to retrieve the backup samples safely stored in Svalbard and to reestablish the Syrian seed collection in partner institutes in Morocco and Lebanon. This "rescue operation" (ibid.) averted an irreplaceable loss of agricultural biodiversity that would have been inescapable without the support of the Svalbard Global Seed Vault. It is a poignant example of how the availability of natural resources for food production, and hence the question of food security, is deeply entangled in a co-constitutive web of global and local ecological, political, social, and technological conditions and changes.

The Svalbard Global Seed Vault, one of the decisive factors in the success story of saving the Syrian agrobiodiversity collection from extinction, is a safety bunker for seeds built deep into a permafrost mountain on the Arctic island of Spitsbergen in the Norwegian Svalbard archipelago, which lies about 4,800 kilometres north of Aleppo, 2,000 kilometres north of Oslo, and a little more than 1,300 kilometres south of the North Pole. The Seed Vault was opened in 2008 and is run by a tripartite partnership between the Norwegian Government, the Crop Trust based in Germany, and the Nordic Genetic Resource Center (NordGen) based in Sweden. Stories about a "doomsday vault" in the Arctic saving the seeds to feed a post-apocalyptic world (e.g. Nature 2008; The Guardian 2015; Heatherington/Perley 2017), together with spectacular images of a concrete entrance into a snow-covered mountain radiating a turquoise light into the dark of the Arctic winter, have earned the Seed Vault global attention. Behind the mysterious portal that separates the outside world from the interior of the vault through a security door, a tunnel leads 120 metres deep into the mountain and into what has been called "the most biodiverse room in the world" (Wickson 2016: 788). As of July 2025, more than 1.3 million seed samples of over six thousand species from all around the world (NordGen n.d. b) are conserved in airtight packages sealed in boxes stored on shelves in currently only one of three storage halls, each with a capacity to store a total of 1.5 million seed samples. The seeds are preserved in suspended animation, frozen at -18 degrees Celsius using a technological refrigeration system. They are further protected by the Arctic permafrost surrounding the vault, which guarantees a year-round temperature of no higher than -3 degrees Celsius. The Seed Vault's purpose and promise is to conserve these seeds "in perpetuity" (Fowler 2016: 18) as a safety backup for the world's endangered biodiversity for food and agriculture.

In the social sciences and humanities, the Svalbard Global Seed Vault has attracted attention as a technoscientific promise of security and futurity in the face of increasingly unpredictable and alarming social-ecological futures.[1] Scholars from a range of disciplines including sociology, cultural studies, and philosophy as well as interdisciplinary research fields such as science and technology studies (STS), heritage studies, and the environmental humanities, have analysed the social, cultural, and political implications of the Seed Vault's efforts to conserve endangered biodiversity for food and agriculture. Their analyses can be divided into two groups with one focusing on *what* the Seed Vault conserves and the other asking *what for*. The first group foregrounds *the seeds* conserved in the Svalbard Global Seed Vault. Asking what they represent, these authors explore the premises and implications of the Seed Vault's approach to agrobiodiversity conservation. They examine how the Seed Vault operates within a bio-reductionist (van Dooren 2009a, 2009b), molecular (Wickson 2016; Wolff 2020), and information paradigm (Alpsancar 2016), all of which allow the Seed Vault and genebanks more generally to turn agrobiodiversity into accumulable biocapital (Breithoff/Harrison 2020; Dalyan 2021; Paajanen 2024) while disregarding or even perpetuating

1 The term "social-ecological" encapsulates an argument in environmental sociology that the social and the ecological are entangled in multiple ways rather than denoting separate spheres. While views differ as to how exactly the two are intertwined, in this book I argue for an understanding of social-ecological realities as co-constitutive, co-emergent, and interdependent.

the ongoing destruction of agrobiodiversity outside of seed banks (Nadim 2018). The second group of accounts focuses on *the promises* the Seed Vault embodies in light of the social-ecological threats and challenges that come with global agrobiodiversity loss and climate change. These authors analyse it as a social or political technology of salvation (Fava 2013), securitisation (Wolff 2020), or "becoming-safe" through both (Boschen 2022: 112–153), as well as a technoscientific means of generating specific forms of hope, security, and futurity (Harrison 2017; Harrison et al. 2020: 51–140).

The state of research on the Svalbard Global Seed Vault offers a multi-faceted and differentiated understanding of its premises, promises, and (bio)politics. However, most of these accounts isolate the Seed Vault from many of the contexts and relations it is entangled in and which are crucial for a comprehensive assessment. For one thing, they pay little attention to the global system of agrobiodiversity conservation the Seed Vault is embedded in, which involves far more than the storage of seeds on ice, and to the ways in which the history of this global system matters to the story of the Seed Vault. This lack of engagement with the history of seed conservation beyond the Seed Vault is accompanied by a lack of attention to the historicity of not only seed conservation but also agrobiodiversity loss, which has only become a global ecological problem in the context of twentieth century agricultural modernisation. Secondly, by discussing the Seed Vault as a technology or set of technologies handling a *diversity of seeds*, most studies disregard the *diversity of actors* the Seed Vault assembles along with seeds and the ways in which they co-shape its conservation efforts and politics. Put differently, inasmuch as these analyses foreground the more-than-human (bio)politics of the Seed Vault, its more-than-*non*human politics fade into the background. My study responds to these desiderata by focusing attention on the Seed Vault's worldly entanglements. In doing so, I highlight ambiguities and spaces of indeterminacy in the story of the Seed Vault, as well as its potential to serve and conserve more-than-human relations other than the biocapitalist and biocolonial ones it has emerged from.

A first important contextualisation of the Seed Vault is that it is not a genebank in the proper sense, but a backup repository for the secure storage of safety duplicates of seed collections held in other genebanks all around the world. These support the work of farmers, breeders, and scientists by making seeds and the genetic traits they carry available for agricultural practice, research, and development. Whereas such genebanks have what STS scholars Pablo A. Pellegrini and Galo E. Balatti (2016: 2761–2763) call an "assistentialist" or a "productivist profile", meaning that they either assist farmers directly or focus on research and breeding, the Seed Vault is a strictly "preservationist" (Pellegrini/Balatti 2016: 2763–2764) facility focused on the long-term storage of a broad range of biodiversity. The conservation of the diversity of plants and varieties for food and agriculture in *ex situ* conservation environments – that is, outside of plants' natural habitats – has gained increasing importance in global agricultural practice since the mid-twentieth century in the wake of an enormous loss of crop diversity *in situ* – meaning the diversity cultivated in farmers' fields. According to broad estimates by the United Nations' Food and Agriculture Organization (FAO), the diversity of crops cultivated worldwide declined by about 75 per cent over the course of the twentieth century (see Khoury et al. 2022). According to the FAO's latest report on *The State of the World's Biodiversity for Food and Agriculture*, less than two hundred out of around six thousand currently cultivated plant species substantially contribute to global food output, and as few as nine of these account for 66 per cent of total global crop production (FAO 2019: 114).[2]

Historical research has shown that this diminishment of agricultural biodiversity in farmers' fields is deeply entangled with the modernisation and globalisation of agricultural production around the world in the second half

2 There are an enormous number of studies that quantify the diversity of plants existing in the world and the decline of crop diversity in different ways, based on a variety of metrics, indicators, and geographical as well as temporal scales. For a literature review see Khoury et al. (2022). Of course, the loss of biological diversity exceeds the loss of crop diversity that is of interest in this book. The rate of biological diversity loss throughout the twentieth century is so great – and much higher than the expected background extinction rate – that scientists now largely agree that the earth is heading towards a Sixth Mass Extinction (see Dirzo/Raven 2003; Barnosky et al. 2011; Ceballos et al. 2015; Cowie/Bouchet/Fontaine 2022). The previous five mass extinction events in the Earth's history, which date back approximately 65 to 443 million years, are attributed to rapid climate, atmospheric, and biochemical changes caused by asteroid impacts, tectonic, or volcanic activity. In contrast, the anticipated sixth mass extinction is mainly ascribed to anthropogenic causal forces. A significant driver is modern agriculture, which destroys ecological habitats and conditions for biological diversity to thrive through unrelenting land reclamation, non-replenishable resource overexploitation, excessive use of agrochemicals, as well as through the de-cultivation of biological diversity in farmers' fields (see Böhning-Gaese/Bauer 2022; Khoury et al. 2022). This dual implication of agriculture in biodiversity loss, for which it is partly responsible and by which it is affected, is particularly evident in the loss of agricultural biodiversity, as I unfold throughout this book.

of the twentieth century, which largely relied on homogenised monocropping and the large-scale introduction of plant breeds producing high yields through the use of agrochemicals (see Flitner/Heins 2002; Kloppenburg [1988] 2004; Saraiva 2013; Fenzi/Bonneuil 2016; Peres 2016; Curry 2017a, 2022b). Although the social-ecological repercussions of this form of global agricultural production became widely evident as early as the 1960s, most of the international agro-political community was convinced that the high productivity and efficiency of globalised modern agriculture made this system indispensable if the entire world population was to be fed. Loss of crop diversity in farmers' fields was seen as an undesirable but unavoidable side effect. It was in this context that *ex situ* conservation of crop diversity in genebanks became the primary international strategy against agrobiodiversity loss.

Today, the agricultural biodiversity conserved in *ex situ* conservation facilities is much greater than the diversity currently cultivated *in situ*, and many of these facilities steward a unique botanical and genetic diversity that would be irretrievable if lost. The problem with this, as history has shown, is that seed and genebanks are exposed to a variety of threats ranging from man-made disasters such as wars to hurricanes and floods, and also to more mundane threats such as financial and capacity limits or technological problems (Fowler 2008c: 12, 2016: 90–93). The case of the Syrian national genebank is only one in a long series of examples of seed collections that have been lost as a result of such events, in most cases irretrievably (see Fu 2017). The Svalbard Global Seed Vault is a direct response to this endangerment of biodiversity for food and agriculture both *in situ* and *ex situ*. It was created to provide a storage space that was as secure as possible, so that genebanks around the world could conserve safety duplicates of their seed collections for as long as possible. By preserving for the long term what the founder of the Seed Vault, Cary Fowler, refers to as "the biological foundation of agriculture and arguably humankind's most important natural resource" (2016: 18), the Seed Vault promises to be able to undo local losses of diversity, enable breeders to continuously adapt agricultural production around the world to changing environmental and climate conditions, and thus ensure future food security.

The seeds currently conserved in Svalbard come from 129 seed and genebanks all over the world (NordGen n.d. b) and represent a wide range of farmers' varieties, crop breeds, and wild plants relevant for food and agriculture (Norwegian Ministry of Agriculture and Food n.d. b). The Seed Vault stores these seeds under a "'black box' agreement" (Fowler 2008b: 190; see also Asdal 2025: 50), meaning that every deposit remains the sole property of the depositor at all times. Accordingly, the Seed Vault also differs from other genebanks in that it has no authority over the collection of seeds it stores. In this sense, it is not an international agrobiodiversity collection in the hands of the Norwegian government or the international actors responsible for its operation, but rather a manifestation and nodal point of previous and ongoing multilateral cooperation. This cooperation, in turn, exceeds activities related to the storage of seeds in Svalbard insofar as it also encompasses the continuous development and negotiation of the technoscientific, political, social-ecological, and other conditions under which it takes place. What is more, all cooperating partners bring a set of concerns, interests, needs, knowledges, histories, and more to the table, thus entangling their respective worlds in the cooperation and its outcomes.

Investigating the Seed Vault as a nodal point of a global network of genebanks and seed flows, but also of a diversity of actors, practices, imaginaries, knowledges, stories, histories, and more, in the interplay of which past, present, and future more-than-human and more-than-nonhuman relations and realities are being negotiated, is the second crucial contextualisation of the Seed Vault's premises, promises, and politics that my study performs. My inquiry into the world of agrobiodiversity conservation through the Svalbard Global Seed Vault complements and expands existing research in the social sciences and humanities by situating the Seed Vault within its global and historical, material-discursive and political entanglements while also cultivating attention to ambiguities, indeterminacies, and unexpected collaborations.

The research approach underlying this study is grounded in feminist technoscience studies. This transdisciplinary field, which emerged in the 1980s, constitutes a particular strand of both feminist theory and science (and technology) studies that investigates "the sociocultural embeddedness of all scientific and technological theories and practices" (Åsberg/Lykke 2010: 299).[3] The notion of technoscience encapsulates a basic assumption and central critique of this research field: by conjoining "apprehension and intervention" (Murphy 2012: 71) in a single term, it challenges the idea of an independence of scientific knowledge production from its 'good' or 'bad' technological application. Instead, feminist technoscience studies scholars argue that scientific knowledge production is inseparable from technological intervention and instrumentality insofar as it is "not mere observation, it [is] *world making* as

3 For an overview of the field and the diverse range of scholars who have shaped and continue to shape it, see Åsberg and Lykke (2010); Catalyst Journal Editorial Board (2015).

the very scientific acts of study involve[] altering the world" (Murphy 2012: 71, emph. added). A key strand within feminist technoscience studies that is particularly relevant for this book is the work of Donna Haraway (1988, 1997, 2016). Arguing that "[it] matters what worlds world worlds" (2016: 35), Haraway emphasises that it is crucial for feminist technoscience studies to attend not only to technoscientific world making practices, but also to the worlds these practices are situated in.

Following an earlier version of this argument in Haraway's work – her claim that "[n]othing comes without its world, so trying to know those worlds is crucial" (1997: 37) – I propose to consider the Seed Vault as a space that assembles not only a diversity of seeds, but also a diversity of worlds that come with seeds and other concerns related to the Seed Vault. This makes it possible to develop a differentiated understanding of the ways in which the Seed Vault is both the product of historical social-ecological processes and political struggles and an agentic force in ongoing material-discursive processes and struggles of social-ecological becoming. It is in this sense that the analysis I unfold throughout the book – as the book's subtitle indicates – is not an account *of* the Seed Vault so much as an inquiry into the larger world of agrobiodiversity conservation *through* the Seed Vault. Conceiving and introducing the Seed Vault as a nodal point of agrobiodiversity conservation efforts, which primarily, but not exclusively, relates to international *ex situ* conservation efforts, I show that by collecting a diversity of seeds from all over the world it has created a technoscientific and political arena assembling a diversity of actors, practices, discourses, histories, imaginaries, and more. It is in the interplay of these that worlds come to matter through the Svalbard Global Seed Vault.

The notion of the world is an unusual category to think with for a sociologist. According to Alexandra Schauer (2023), who examines the loss of meaning of the world in late modernity, the world has traditionally not been considered suitable as a basic sociological concept because it denotes "an 'absolute metaphor' referring to a totality that is not tangible as such" (Schauer 2023: 16, quoting Blumenberg [1960] 2009: 10, author's translation; see also Blumenberg [1960] 2010: 3).[4] While this observation is empirically apt, the explanation hinges on a canonical conception of the world as a "one-world

4 According to Schauer (ibid.), scholarly interest in the notion of the world is largely restricted to phenomenological discussions about worldliness as part of the human condition (the Heideggerian *In-der-Welt-sein*) and associated socio-philosophical debates

world" (Law 2015). For some decades now, anthropologists have been criticising such a monolithic understanding of the world for omitting alternative and more pluralistic theorisations in other bodies of knowledge that are excluded from the canon of the dominant modern social sciences and humanities (e.g. Viveiros de Castro 1998, 2014; Descola 2005, 2014). Drawing on the anthropological research of Eduardo Viveiros de Castro, therefore, Bruno Latour has reproached sociology with "anthropology blindness" (2004: 453) and "ethnocentrism" (2004: 455): "For the sociologist, nature, the world, the cosmos, is simply there; and since humans share basic characteristics, our view of the world is, at baseline, the same everywhere" (Latour 2004: 453).[5] Where there are different worldviews, he argues, sociology usually considers them as different perspectives on one and the same world – manifestations of multiple cultures differently interpreting a singular nature, world, or universe. Taking anthropological accounts of fundamentally different ways of knowing and being in the world into account and drawing on William James' conception of the world as a "pluriverse" (2004: 454), Latour argues that sociology's adherence to the idea that there is only one nature, one cosmos, and one world is no longer tenable.[6]

Echoing this critique of the Eurocentric, modernist idea of "mononaturalism and multiculturalism" (Law 2015: 126), a growing body of anthropo-

---

about what it means, for instance, for humans as opposed to nonhumans to be in the world.

5 The postcolonial sociologist Gurminder K. Bhambra (2014) argues that the separation of sociology from anthropology is premised on a "profound division [...] between the social sciences whose remit is the modern world, and anthropology, whose remit is the traditional" (Bhambra 2014: 418). This separation has historically absolved sociology from the necessity of reflecting on the constitutive dependency of its categories on the construction of an Other and "from any consideration of colonial and imperial entanglements" (ibid.). Bhambra counters this de-responsibilisation by emphasising the importance of considering "the 'connected histories' that are the mutual conditions of emergence of both" (Bhambra 2014: 420).

6 In the context of this critique and following the work of Isabelle (Stengers [1997] 2010, [1997] 2011), Latour foregrounds the notion of the *cosmos* rather than the world. In doing so, he counters "the peace terms" (Latour 2004) of Ulrich Beck's Kantian cosmopolitanism (Beck 2004) with a notion of *cosmopolitics* that recognises the deep entanglement of the human and the nonhuman world and integrates the latter into the sphere of politics. While this argument informs my thinking about the politics of world-making discussed in this book, Latour is less central to my embracing of the notion of the world than the accounts presented in the following.

logical scholarship developed in communication with Indigenous and other marginalised worlds refusing to be obliterated by the universalist ontology of modernity theorises the world as a fractiverse (Law 2015), a multiverse (Verran 2018), a pluriverse (Escobar 2018, 2020), or a world of many worlds, as Marisol de la Cadena and Mario Blaser (2018a) put it following the Zapatistas.[7] These accounts and the worlds they are rooted in consider the idea of a one-world world not as a universal ontological truth but as one particular among many situated and partial ways of knowing and being in the world. They counter the canonical assumption of a singular world merely differently conceived by divergent cultural knowledges with an ontological pluralism that politicises the question of ontology. That is, worldly realities are neither ahistorically nor universally given but constantly emerge in "contingent and relational enactments" (Law 2015: 134). Put differently, the world is not a singular metaphysical being so much as a multiplicity of concrete and ongoing doings (see also Tsing 2015: 21–22).[8] Crucially, drawing attention to heterogeneous practices of world-making giving rise to a multiplicity of co-existing worlds does not mean altogether dismissing the phenomenological and philosophical argument that the world is a hardly tangible totality. Rather, the aim is to contextualise this argument as one particular and situated way of knowing and being in the world (see Law 2015: 128). Ontological pluralism, in other words, "is not a matter of either/or but of *both/and*" (Blaser 2016: 565, emph. added).

7 The Zapatistas are a predominantly Indigenous peasant movement in the state of Chiapas in Southern Mexico. Since the adoption of the North American Free Trade Agreement (NAFTA) between the USA and Mexico in 1994, they have forcefully been defending their world against the Mexican government's neoliberal policy of privatising communal land and deregulating markets. Guiding their struggle is the motto: "The world we want is a world in which many worlds fit" (Ejército Zapatista de Liberación Nacional (EZLN) 1996, as quoted and translated by de la Cadena/Blaser 2018b: 1). For a more in-depth scholarly discussion of the Zapatista movement see, e.g., Mignolo (2011: 213–251) and Klein (2015).

8 It should be noted here that Schauer (2023) counters the sociological aversion to the world as an analytical category with a sociological theory of modern and late modern relationships to the world that builds on Hannah Arendt's conceptualisation of the world as a political phenomenon characterised by historicity, plurality, and malleability. Despite the acknowledgment of plurality, this approach differs from the anthropological understanding of worlds as heterogeneous situated doings outlined here insofar as neither Arendt nor Schauer challenges the metaphysics of a singular world; they both understand plurality in a multicultural or epistemological rather than an ontological sense.

The anthropological body of scholarship developing this understanding of *the world as both a totality and a multiplicity* constitutes the first of three pillars buttressing my thinking about the Svalbard Global Seed Vault as an arena assembling a diversity of worlds.

The second pillar is an understanding of the world(s) as an emergent materialisation of ongoing and heterogeneous doings. Haraway (2008, 2016) has prominently encapsulated this idea in the notion of *worlding*, which articulates a situated, practice-based, processual, and relational understanding of how beings and worlds come to matter. She argues: "Ontologically heterogeneous partners become who and what they are in relational material-semiotic worlding. Natures, cultures, subjects, and objects do not preexist their intertwined worldings." (Haraway 2016: 12–13) The notion of worlding is only the most recent articulation of Haraway's decades-long critical engagement with technoscientific abstractions such as Nature and the World – capitalised to indicate the assumed universal ontology the terms represent. In a seminal article on her approach to the study of nature published over three decades ago, Haraway argued: "If the world exists for us as 'nature,' this designates a kind of relationship, an achievement among many actors, not all of them human, not all of them organic, not all of them technological" (1992: 297). The point she is making here is that the world not only comes to matter in concrete and continuously ongoing practices and is therefore multiple, but that such practices are enactments of specific more-than-human relations.

How we – whoever that we is – represent the world, then, is not the same as what it is; what dominant Western science calls Nature is not a given, objective reality but a particular, situated way of making sense of reality. According to Haraway, "the world is precisely what gets lost in doctrines of representation and scientific objectivity" (1992: 313). In contrast, while her thinking "does not end up with an 'objective world' or 'nature,' [...] it certainly does insist on the *world*" (Haraway 1992: 313, emph. in orig.) as a force that *matters* – meaning that it makes a difference in terms of both meaning and materiality. Haraway's recent condensation of this argument in the succinct phrase that it matters what worlds world worlds (see above) highlights that, if worlds come to matter in ongoing, heterogeneous, material-discursive, more-than-non/human processes and practices of worlding, it matters where, when, how, with whom, to what end, on what grounds, and so forth these processes and practices take place. These are some of the questions that inform my investigation of the many worldings the Svalbard Global Seed Vault assembles and through which it comes to matter.

The third pillar of this focus on the Seed Vault as a nodal point of diverse worlds and worldings is the methodology of Situational Analysis, which has guided the empirical-analytical process underlying my study. Situational Analysis is an explorative, qualitative research approach in the interpretive paradigm developed by Adele E. Clarke and others (Clarke 2005; Clarke/Friese/Washburn 2018) as an extension of Grounded Theory in the tradition of Anselm Strauss (see e.g. Corbin/Strauss 2015). Unlike Grounded Theory, it focuses analytical attention not on individual subjects and bounded social groups as sources of knowledge and action but on relational and co-constitutive material-discursive and more-than-human social ecologies. These are explored by mapping the complex web of sites, individual and collective actors, and material and discursive elements that constitute a situation of inquiry as well as by tracing the relations between them. In doing so, Situational Analysis puts the emphasis on identifying material-discursive arenas of sustained interest and exploring the diversity of *social worlds* these arenas assemble and are shaped by. Social worlds, in Situational Analysis, denote mutable social wholes that consist of the individual and collective actors they comprise, and are held together by shared interests, concerns, practices, discourses, and technologies. Clarke and her colleagues use the terms social worlds and worlds synonymously. However, the notion of social worlds, which is rooted in Strauss' interactionist theory, comes with an anthropocentric bias that I seek to eschew by using only the term "worlds" in the sense of the theorisation of world(ing)s discussed above.

Although Haraway's thinking too is woven into the theoretical premises of Situational Analysis, as can be seen, for example, in the conception of arenas and social worlds as "lively, dynamic, deeply relational entities" (Clarke/Friese/Washburn 2018: 148), Clarke and colleagues consider it "too awkward to say that they are 'worlding' and 'arena-ing'" (Clarke/Friese/Washburn 2018: 148). In contrast to this assessment, I consider Situational Analysis an excellent means to explore how worlds world worlds. To do so, in contrast to previous scholarly accounts of the Svalbard Global Seed Vault that have primarily analysed it as a technology or set of technologies, practices, and rationalities, I propose to conceive the Seed Vault as an arena that assembles a diversity of worlds and worldings that come with a diversity of actors, practices, technologies, discourses, histories, stories, imaginaries, and more. The methodology of Situational Analysis enables this inquiry insofar as it is designed to map all these elements, dimensions, and the relations between them. Its aim is to stay with the complexity of a situation of inquiry rather than zooming in on one dimension (e.g. *either* practices *or* discourses *or* certain institutions) at the expense of both other dimensions and the multidimensionality of the phenomenon under investigation. It provides a conceptual and methodical framework that allows for exploring how phenomena – such as agrobiodiversity loss and conservation – and the arenas in which they are negotiated – such as the Seed Vault – come to matter in the interplay of a multiplicity of worlds and worldings. Thus, it becomes possible to show that the Seed Vault's politics of world-making is not merely a manifestation of the world it has arisen from but comes to matter in the interplay of the many worlds and worldings it assembles. This, in turn, allows me to develop a critical analysis of the Seed Vault that, following the tradition of feminist technoscience studies, enacts a mode of critique that is both deconstructive and productive – something that I consider both fruitful and essential for a sociology that seeks to grapple with the current social-ecological condition.

The empirical foundation of my inquiry into the world(s) and worldings of agrobiodiversity conservation through the Svalbard Global Seed Vault is a multi-modal study composed of ethnographic field research, qualitative expert interviews, and document analysis. By means of an abductive research process, in which empirical and theoretical engagement with the situation of inquiry inform one another in a circular manner, I conducted two ethnographic field research trips to Svalbard and the Global Seed Vault in June 2019 and February 2020 and a short field trip to a German plant research institute and depositor to the Seed Vault in December 2019. I further conducted fourteen semi-structured qualitative expert interviews that took place between May 2019 and March 2021, with one exception in March 2023. The experts I interviewed are members of the partner institutions behind the Seed Vault and other important institutions and organisations in the *ex situ* conservation environment, as well as activists from the non-governmental and civil society sector engaged in agrobiodiversity politics. The analyses I develop based on this body of empirical data are further undergirded by text and film documents that provide deeper insights than those that can be extrapolated from the other empirical material. Most previous social-scientific and humanities research on the Seed Vault has not been based on this kind of in-depth on-site empirical research.[9] My study shows that a thorough empirical engagement

9 Notable exceptions are the works of Harrison et al. (2020) and Boschen (2022). However, other than in my study, in both these cases, the Seed Vault is not the heart of the

based on more than document analysis is essential insofar as many of the publicly available sources of knowledge about the Seed Vault paint an abbreviated and sometimes inaccurate picture. Against this background, the empirically grounded, comprehensive study of the Seed Vault presented here also offers a foundation for further social-scientific and humanities research in the future.

Before concluding this introduction with an outline of the structure of the book, I would like to clarify a few key terms. The conservation of seeds and other plant material in *ex situ* conservation environments targets a specific type of biological diversity referred to as *plant genetic diversity*. According to the United Nations' Convention on Biological Diversity from 1992, biological diversity or in short *biodiversity* comprises the diversity of ecosystems (land- and sea-based) and species (plant, animal, microbial, and other) as well as the diversity of variations within species. Plant genetic diversity refers to the genetic diversity of and within plant species. While *biodiversity loss* comprises the loss of plant genetic diversity, it refers more broadly to the loss of genetic, species, and ecosystemic diversity. To refer specifically to the loss of plant genetic diversity, scientists also use the term *genetic erosion* (see chapter 1.2). This notion was coined in the context of scientific and political debates about the devastating consequences of the loss of *plant genetic diversity for food and agriculture*, which is a specific element of plant genetic diversity. The broader biological diversity relevant to food and agriculture is also called agricultural biodiversity or *agrobiodiversity*. The latter is at the centre of interest in this book, against the backdrop that the *ex situ* conservation efforts of the Svalbard Global Seed Vault and the genebanks whose safety duplicates it stores focus on plant genetic diversity for food and agriculture. Although the more commonly used terms in the context of genebanking are plant genetic diversity or *plant genetic resources for food and agriculture*, throughout this book I frequently use the term agrobiodiversity despite the fact that it encompasses much more than plant genetic resources for food and agriculture. In doing so, I aim to retain awareness of the ontological embeddedness of (even extracted) plant genetic diversity in (social-)ecological contexts and to not uncritically reproduce the *ex situating* logic of the genetic paradigm.

Last but not least, readers will read about genebanks more than about seed banks. The term *seed banks* usually refers to conservation facilities exclusively conserving seeds, such as farmers' and community seed banks, as well as

---

analysis, but one of a number of case studies through which the authors develop an analysis of heritage or seed banking practices more generally.

seed repositories such as the Svalbard Global Seed Vault. The more common term *genebanks* denotes conservation facilities that conserve seeds along with other plant genetic material. The conservation strategies of such genebanks include seed freezing, but also field conservation, *in vitro* conservation, and cryopreservation of plant tissue in liquid nitrogen. Underlying this conservation approach is the conviction that "[s]pecific genes are what make a crop variety resistant to heat and drought, for example, or tolerant of pests and diseases, and are, therefore, the basis of agricultural adaptation" (Crop Trust 2022: par. 4). Insofar as genebanks often supply scientists and breeders rather than farmers, their objective is to conserve the genetic traits embodied in the germplasm of seeds and other plant material – hence the term "genebank" (see chapter 4.1). A sociological discussion of the premises and implications of these key terms is unfolded in the chapters that follow.

Outline of the Book

The first chapter traces the historical emergence of the Svalbard Global Seed Vault by unfolding a short history of agrobiodiversity loss and international conservation efforts from the mid-twentieth century to the present. It begins with the so-called 'Green Revolution' of the post-World War II era, a US-led agricultural development programme with the professed goal of preventing hunger and societal upheaval in the Global South by increasing agricultural production. The chapter shows how the modernisation approach of this programme and the agricultural realities it implemented in many countries around the world spawned the loss of agricultural biodiversity the world now faces as a global social-ecological problem (1.1). Beginning in the 1960s, alarmed voices in the scientific community warned about the devastating consequences of "genetic erosion" (Frankel et al. [1969] 1970; Harlan 1975), which triggered a coordinated international effort in the UN environment to systematically conserve the plant genetic "heritage of mankind" (FAO 1983). I delineate how *ex situ* conservation became the predominant international response to agrobiodiversity loss in this historical-institutional context (1.2). I then trace the ensuing international negotiations over sovereignty claims and property rights, access and benefit-sharing regulations, breeders' and farmers' rights, and other issues (1.3) and discuss how they eventually resulted in a cooperative international agro-political climate favourable to the establishment of the Svalbard Global Seed Vault as the first international agrobiodiversity conservation facility (1.4). I conclude the first chapter by delineating the conditions of possibility for the establishment of the Seed Vault as well as its technical and operational functioning (1.5).

Chapter two begins by developing the approach of my inquiry into the Seed Vault and the world(ing)s of agrobiodiversity conservation through an engagement with the state of social-scientific and other relevant research. I discuss in more detail how scholars in the social sciences and humanities have analysed and theorised the work of the Seed Vault so far, and elaborate on other research that informs my thinking. My approach comprises engagements with the science, technologies, and politics of agriculture and (agro)biodiversity conservation beyond the Seed Vault in historical scholarship as well as in STS and the environmental humanities (2.1). The second section elucidates the methodological and empirical foundations of my study. I introduce the methodology of Situational Analysis, outlining the theoretical and methodological premises, concepts, and ideas that are relevant to my research. I then go on to unfold the empirical-analytical process, presenting sites, phases, and methods of my study and explaining some of the empirical choices made and limits faced (2.2).

Chapters three to five take the reader on my empirical journey to and beyond Svalbard and the Global Seed Vault, unfolding the different ways in which it matters what histories, imaginaries, and more-than-human relations, what worlds and worldings make the Seed Vault and the phenomena of agrobiodiversity loss and conservation. Divided into two sections each, all three chapters develop a critical analysis that deconstructs and constructively juxtaposes the Seed Vault and the *ex situ* world of conservation with other worlds and worldings the vault assembles around the issue of agrobiodiversity loss. This type of critical analysis that is both deconstructive and productive reflects the concern of feminist technoscience studies not only with the implications and problems of hegemonic technoscientific projects, but also with "forms of technoscientific counter-conduct" (Murphy 2012: 72). Crucially, the aim of placing emphasis on the latter is not to renounce technoscience or compare 'bad' (e.g. because it reproduces established power relations) with 'good' technoscience (e.g. one that is more democratic). Rather, the aim is to highlight the diversity of forms of world-making, their entanglements and interdependencies, as well as the transformative potential of interventions into dominant technoscience.

Chapter three follows the paths of my first journey to Svalbard. Reiterating Fowler's assertion that "[t]o understand the Seed Vault, you first need a feeling for its home, Svalbard" (2016: 25), I begin by situating the Seed Vault in Svalbard. Svalbard typically matters in stories of the Seed Vault with regard to its cold temperatures, remoteness, and the assumed geological and political stability of the archipelago, which are seen to make it the safest possible location for the Seed Vault. In contradistinction to this, I focus on how Svalbard gives the Seed Vault a more-than-*non*human home. To do so, I situate the Seed Vault in the history of human activity on the archipelago as told by the Svalbard Museum in Longyearbyen – from its colonisation in the age of empires through different phases of extractivist resource depletion to the present day. Reading this history as a history of divergent iterations of an extractivist and frontierist relation to the natural/ised world, I trace its continuities as they come to matter in the Seed Vault and *ex situ* agrobiodiversity conservation (3.1).

The second section shifts the focus from the histories embodied in the Seed Vault to the worldly relations and politics it enacts in the present. Recounting what it was envisioned and designed to be by the people and organisations behind it, I elaborate the logic of backup security that characterises the work of the Seed Vault. Only seven years after it was established, this logic was undermined by a water intrusion into the Seed Vault due to melting permafrost. While the media narrative foregrounds the irony of a climate change-related threat to the stronghold against climate change, my interview material shows that this story misses some important aspects. It also shows that the response of the partners behind the Seed Vault to this incident – a time- and resource-consuming reconstruction of the facility – is characterised by the same logic of security-through-backup as the Seed Vault's work of safety duplication. The chapter closes by discussing how the Seed Vault and its specific rationality of conservation is received by conservationists and activists in the world of agrobiodiversity conservation. It is through these accounts that I finally unfold the analytical approach to the Seed Vault I propose and have introduced above, namely to conceive it as a political arena assembling a diversity of worlds and forms of world-making (3.2).

Chapter four extends the focus from Svalbard and the Global Seed Vault to the worlds and worldings that come with the seeds conserved inside the vault. It brings into view how different more-than-human relations come with divergent forms of agrobiodiversity conservation. I first introduce what I call, following one of my interview partners, the *ex situ* world of conservation, which refers to more than the global network of genebanks whose collections the Seed Vault reinsures. By elaborating on the work of *ex situ* conservation in genebanks and its relation to *in situ* and on-farm conservation approaches, I develop an analysis of "the *ex situ* world" as a set of practices, rationalities, technologies, and more-than-human relations. It is shaped, among other things, by an *ex situ* worldview that is prone to neglect the perils of biological and agricultural de-diversification outside the genebank insofar as at its heart, it is characterised by an understanding of vegetal life as a composition of extractible resources (4.1).

The second section traces the interdependencies of the *ex situ* world of conservation with other worlds and modes of worlding. It begins by reviewing how this matter is discussed in the *ex situ* world of conservation, where international interdependency in access to plant genetic diversity is becoming a matter of growing concern in light of the social-ecological challenges of the present. I then discuss the question of global interdependence through the Parque de la Papa (English: Potato Park) in the Peruvian Andes, which lies in one of the world's primary centres of biodiversity and is the first Indigenous organisation ever to have put heritage seeds into storage in the Svalbard Global Seed Vault.[10] The Potato Park is an interesting case in several respects. Since it first deposited seeds in Svalbard in 2015, members of the park have cultivated a collaboration with the *ex situ* world that is oriented towards securely preserving the unique biocultural heritage cultivated and conserved in their territory as well as towards a mutually beneficial transformation of global food systems (see Khoury et al. 2019; Argumedo et al. 2021). This collaboration is noteworthy insofar as the Potato Park comes with a world that fundamentally differs from the *ex situ* world. The conservation practices and rationalities of this world are situated within a deeply relational cosmology in which more-than-human relations are relations of kinship and mutual care.[11] Taking this cosmological otherness seriously in the discussion of global interdependence through the

10 In referring to the Potato Park and its inhabitants as "Indigenous", I follow their self-description on the internet, in publications about the park, as well as in the interview I had the chance to conduct with one of the park's founders. Throughout this book, I capitalise the term Indigenous to indicate that it is not a substantial category but a historically specific and political term. On the notion of "Indigeneity", see chapter 4.2, footnote 20.

11 I use the term "cosmology" as it is understood in anthropology (e.g. Viveiros de Castro 2015; de la Cadena/Blaser 2018b; Verran 2018) rather than in physics or philosophy. In these disciplines *cosmology* usually refers to a universal cosmos whereas anthropologists such as those quoted here deploy a pluralist understanding of *cosmologies* as situated and partial knowledges of the world understood as a multiverse (Verran 2018) or of one particular world among many worlds in a pluriverse (de la Cadena/Blaser 2018b). Differing and incommensurable cosmologies, from this perspective, cannot be traced back "to imaginary ways of 'seeing the world' but rather to the real worlds that are being seen" (Viveiros de Castro 2015: 64; see also Blaser 2016).

Potato Park, I develop a more-than-human understanding of *earthly interdependence* as something that exceeds resource and food security. This dimension of interdependency affects not only Indigenous and other societies with a relational cosmology but all life on earth, and it is therefore crucial to take this into account when engaging with the social-ecological threats and challenges of the present (4.2).

Chapter five returns to Svalbard and turns to the stories and imaginaries of agrobiodiversity loss and conservation circulating around the Seed Vault, discussing how "[it] matters what stories make worlds [and] what worlds make stories" (Haraway 2016: 12). The first section addresses the popular and persistent, if not unchallenged, biblical narrative portraying the Seed Vault as a "doomsday vault" or a "Noah's Ark" for the world's seeds. Although the partners behind the Seed Vault constantly strive to dispel this narrative because it misconstrues the vault's this-worldly mission, it captures a salvationism that I argue runs through their promissory re/presentation of the Seed Vault. To unfold this argument, I take the reader into a seed storing ritual performed on the occasion of an anniversary event in Svalbard I attended as a participant observer. I discuss the ways in which the religio-symbolically saturated mise-en-scène of the Seed Vault in the ceremony performatively entangles the promissory techno-optimism of the *ex situ* world in biblical narratives of salvation, thus co-producing the story of the Seed Vault's mission as a messianic one (5.1).

The second section explores the Seed Vault's entanglement in other forms of dealing with the loss of agricultural and biological diversity beyond the techno-optimist conservation of seeds in genebanks. I draw attention to one of the actors less evidently assembled by the Seed Vault, an art project I came across during my first field trip to Svalbard. This project consisted of an exhibition of artworks engaging with agrobiocultural diversity, on the one hand, and a performative ritual of burying these artworks for long-term conservation in the permafrost alongside the Seed Vault, on the other hand. The art project provides an intriguing contrast to the engagement with agrobiodiversity loss in the *ex situ* world of conservation and the Seed Vault insofar as it challenges both the *what* and the *how* of *ex situ* conservation. Unlike the techno-optimistic and salvationist seed storing ritual, the ritual of burying artworks in the permafrost was characterised by an affective atmosphere of grief and practices of mourning over social-ecological losses that invite us to rethink what it means to deal and learn to live with past, present, and future loss of agrobiodiversity (5.2).

The book closes with a sixth and final chapter that summarises and reflects on the results of my inquiry into the world of agrobiodiversity conservation through the Svalbard Global Seed Vault. I conclude that a sociology that aims to grapple with the social-ecological catastrophes and transformations of the present and future in a productively critical way needs to cultivate attention to the more-than-human relations underlying both the crises facing the world and the solutions or responses developed to deal with them. The case of the Svalbard Global Seed Vault and *ex situ* agrobiodiversity conservation shows that understanding contemporary technoscientific politics of world- and future-making requires attending not only to what technologies and technoscientific practices do, but also to the worlds and modes of world-making they are embedded in and committed to as well as to the rationalities, imaginaries, and more-than-human relations they come with.

1. A Short History of Agrobiodiversity Loss and Conservation

Farmers have collected and conserved seeds for exchange and future sowing for as long as agriculture has existed. Centralised seed collections conserving seeds of different origins for the purposes of research, breeding, and profit-making, are the forerunners of today's seed banks and their history dates back to the European explorerism of the colonial era. Historians of colonial botany have shown that prospecting for botanical resources all around the world – seeds, plants, as well as knowledge about their cultivation –, introducing them into botanical gardens at home, and transplanting them to plantations in colonial satellite states was an important part of colonial empire-making (see Crosby 1972; Brockway 2002, 2011; Philip 2004; Schiebinger 2004; Schiebinger/Swan 2005; Subramaniam 2024). With the rise of genetics in the early twentieth century, the practices and purposes of collecting, introducing, and conserving seeds and plants fundamentally changed. An often-cited key figure in this development was the Russian botanist and plant geneticist Nikolai I. Vavilov (1887–1943). In the 1920s and 1930s, Vavilov headed over 40 plant exploration and collection expeditions around the world, in the course of which he assembled the largest seed collection of his day and developed his pioneering theory on the "centers of origin of cultivated plants" (Vavilov [1926] 1992a). These centres were areas particularly rich in plant diversity, which Vavilov identified as the genetic places of origin of today's domesticated crops. Most of them were located in what would today be called the Global South.[1]

1 According to Vavilov's "genetic geography" (Kloppenburg [1988] 2004: 175; Flitner 2003), all seven centres of origin he had identified by 1940 lie in what he referred to as the South (see Vavilov 1992b). Jack Harlan (1988: 357) remarks that with six of them located north of the equator, the distinction of North and South must be understood politically, corresponding roughly to what used to be called developed and develop-

Inspired by Vavilov and in the wake of the proliferation of the science of (plant) genetics, more scientists, especially in Germany and the United States soon followed his lead in collecting and conserving the world's "plant resources", as Vavilov referred to them.[2] Insofar as this endeavour was guided by a vision of geneticised agricultural modernisation and hence oriented towards research and breeding, the collecting missions in the first half of the twentieth century came with the establishment of large repositories to store and conserve the collections for the medium term. Thus arose the first seed banks of the kind known today, which were instituted as key elements of a scientified and soon to be commercialised agriculture (see also Fowler 1994: 58–63).

The arc of the history of international agrobiodiversity conservation efforts I recount in this chapter begins in the aftermath of these developments, in the mid-twentieth century, when the loss of biodiversity in agriculture first became a matter of international concern. It ends with the establishment of the Svalbard Global Seed Vault in 2008. Subsequent developments of relevance to the story of the Seed Vault are discussed where they are relevant throughout the chapters that follow. At the starting point of the story I tell here, therefore, seed banks already existed as an integral part of agricultural research and practice. This is a significant factor in how *ex situ* conservation became the core strategy in internationally concerted efforts to conserve agricultural biodiversity. The number of scholars who have paid systematic attention to the history of seed banking is relatively small, although there has been a surge of interest in recent years (see Kloppenburg 1988, [1988] 2004; Flitner 1995; Flitner/Görg/Heins 1998; Flitner/Heins 2002; Flitner 2003; Saraiva 2013; Fenzi/Bonneuil 2016; Peres 2016; Fullilove 2017; Bonneuil 2019; Curry 2017a, 2019a, 2022a, 2022b). These accounts develop intriguing analyses of the history of seed banking as a manifestation and motor of biopolitical empire- and nation-making, modernisa-

ing countries. In what follows, I use the terms "Global South" and "Global North" as they are commonly used today. Countries of the Global South are referred to as 'developing countries' only in the case of direct or indirect quotes from historical accounts discussed. Putting the term in scare quotes serves to mark it as an expression of the Eurocentric developmentalist discourse that I do not mean to reproduce (for a critical account, see Kothari et al. 2019).

2 For detailed accounts of Vavilov's story, work, and impact, see Flitner (2003), Saraiva (2013), Bonneuil (2019), and Curry (2022a).

tion, and globalisation through scientification, geneticisation, securitisation, and other socio-historical developments.[3]

The short history of international agrobiodiversity conservation developed in this chapter builds on these historical accounts; however, it does not offer a detailed reconstruction of their analytical arguments. The aim of what follows is not to provide a comprehensive analysis of all technoscientific, political, economic, social-ecological, and other developments of relative historical importance to agrobiodiversity loss and conservation in general. Rather, it is to trace the specific historical developments leading towards the establishment of the Svalbard Global Seed Vault as the first international backup seed repository for the long-term conservation of the world's agrobiodiversity. To do so, next to historical accounts, I assemble a number of sources from within the world of agrobiodiversity conservation which illuminate the controversies, cooperations, and institutionalisations that were crucial in paving the way for the creation of the Svalbard Global Seed Vault. This tracing of the history of the world of agrobiodiversity conservation through its own accounts provides an initial (historical) mapping.[4]

Section 1.1 begins by elaborating how the development-policy-driven technoscientific modernisation of agriculture in the aftermath of World War II engendered a large-scale loss of agricultural biodiversity, which was officially recognised as a global ecological problem in the mid-1960s. Section 1.2 traces the discussion of "genetic erosion", as this problem was termed, in the international agro-political arena created by the United Nations at that time. It elucidates the historical context in which the conservation of plant seeds in genebanks became the international conservation strategy of choice in the 1970s. In the 1980s, the collection and centralised conservation of the plant genetic "heritage of mankind" became the subject of a highly politicised controversy around access and sovereignty rights between the resource-rich countries of the Global South and the emerging agro-industry of the "gene-

3 I provide a short overview of these accounts in chapter 2.1 and unfold their arguments in greater detail where they are directly relevant to my own arguments.

4 An important source for developments up until 1990 is Cary Fowler and Pat R. Mooney's book *Shattering: Food, Politics, and the Loss of Genetic Diversity* (1990). These authors were two of the first activists who red-flagged the large-scale loss and destruction of plant (genetic) diversity, and both of them have remained active within the field to this day. While Mooney spent his entire career in the civil society sector, Fowler was active in a number of the international institutions introduced in this chapter and eventually became the founder of the Svalbard Global Seed Vault.

poor" Global North. Section 1.3 discusses how this politicised climate influenced the negotiations of the first International Undertaking on Plant Genetic Resources (1983) in the United Nations' Food and Agriculture Organization (FAO) and derailed a first attempt to establish an international seed storage facility in the Arctic. With the UN Convention on Biological Diversity (1992), the landscape of global conservation efforts changed. Section 1.4 elucidates how new commitments, combined with a lack of attention to the agricultural sector's specific needs, engendered new forms of cooperation. These resulted in the first legally binding International Treaty on Plant Genetic Resources for Food and Agriculture (2001), which finally paved the way for the establishment of the Svalbard Global Seed Vault in 2008. Section 1.5 introduces the Seed Vault in organisational and technical terms.

1.1 The "Green Revolution": Agricultural Modernisation and the Making of Agrobiodiversity Loss as a Global Ecological Problem

In the wake of World War II, countries all around the world were faced with severe hunger crises and concomitant societal upheavals. In 1941, the US government under Franklin D. Roosevelt launched an international agricultural development programme to fight hunger in the Global South, especially in Central and South America and Asia. The programme that came to be known as the "Green Revolution" (see e.g. Baranski 2022) was designed by Vice President Henry A. Wallace, a former long-time minister of agriculture and founder of the Hi-Bred Corn Company, one of the large corporations in the US seed industry. Funding primarily came from the Rockefeller Foundation. The pivotal elements of this development initiative were newly established international agricultural research centres (IARCs) oriented towards increasing agricultural production and thus feeding people as fast and directly as possible. The first IARC was established in Mexico as part of the "Mexican Agricultural Program" (Fowler/Mooney 1990: 56–57; Flitner 1995: 131–132; Curry 2017b).[5] A core component of the Green Revolution strategy to increase agricultural production

5 In 1966, this research and plant breeding centre became the International Maize and Wheat Improvement Center (CIMMYT), which still exists today as one of fifteen IARCs worldwide under the auspices of the Consultative Group on International Agricultural Research (CGIAR).

with the help of the IARCs was the breeding of so-called "high-yielding varieties" (Fowler/Mooney 1990: 55) of a region's most important crop plants. These "miracle" (Gan 2017) crop breeds were "high-yield, quick-growth crops with harvest times dramatically accelerated by new inputs that included chemical fertilizers, insecticides, and mechanical equipment, as well as new infrastructure that included irrigation networks, agricultural extension agencies, and rural village banking" (Gan 2017: 62). Insofar as their 'miracle' traits were the result of a high responsiveness to chemicals and irrigation, critics also refer to them as "high-response varieties" (Fowler/Mooney 1990: 54–60; Shiva et al. 2002).[6]

Apart from chemicals and infrastructure, a prerequisite for breeding new varieties is a stock of existing varieties whose desired traits can be harnessed for crop improvement. Therefore, part of the IARCs' everyday business was to collect the greatest possible diversity of locally adapted crop varieties from farmers' fields, which are also called "traditional varieties" or "landraces" (FAO 2019: 11).[7] A corollary of the extensive introduction of 'improved' varieties that would eventually prove to be fatal was the decline of genetic diversity in the field. For the time being, however, the "erosion of landraces was perceived as a side effect of the desired genetic modernization and homogenization of agriculture" (Fenzi/Bonneuil 2016: 75), that is, as a necessary and therefore inevitable consequence of the rationalisation of industrial production. Inasmuch as it made large-scale agricultural production more efficient and served to centralise control of agricultural goods such as seeds, this modernisation was also in the interest of free global trade – a professed goal of US-American and international (agricultural) politics at that time. In 1943, President

6 According to Fowler and Mooney, these developments in crop breeding marked the beginning of a more extensive and lasting "connection between seeds and chemicals" (1990: 130; see below). In 1970, the American agronomist Norman Borlaug, a key figure in most histories of the Green Revolution, received the Nobel Peace Prize for his development of semi-dwarf wheat varieties at CIMMYT. A different challenge to the story of "miracle varieties" than that dismantling them as highly responsive to chemicals can be found in Baranski (2022). She points out that in India, state-supported establishment of irrigation infrastructure actually resulted in "a higher annual growth rate in food crop yields *before* the introduction of new seeds" (Baranski 2022: 7, emph. in orig.).

7 Gan (2019) insightfully discusses the racialised genealogies, implications, and performativity of the taxonomic distinction between "'modern' or 'elite' seeds and genetically modified or 'improved hybrids'; 'traditional' or 'native' landraces; 'weedy' relatives; and 'wild' ancestors" (Gan 2019: 2).

Roosevelt convened an international conference in Hot Springs, Virginia (USA), with the goal of reducing international trade restrictions and advance multilateral cooperation in food and agricultural issues among the Allies and future member states of the United Nations (UN). This conference created a commission that became the FAO in 1945 (Fenzi/Bonneuil 2016: 75; Flitner 1995: 133–134). The FAO remains one of the most important organisations in global agricultural politics to this day.

The Green Revolution strategy strongly shaped international agricultural politics in the UN environment in the aftermath of World War II and must be situated in the historical and socio-political context of that time. While the professed humanitarian impetus of the Green Revolution was to avert hunger crises, critics argue that the Green Revolution strategy also served the US Government as part of its strategy to contain the perceived danger of "Red Revolutions" (Fowler/Mooney 1990: 56; Heins/Flitner 1998: 20), that is, of a proliferation of communism. In Mexico, where the programme was first implemented, extensive agrarian reforms had disempowered large landholders in previous years and peasant farmers were striving for further land reforms. In South and Southeast Asia, where there was also a struggle between the capitalist West and the communist East, it was increasingly recognised that "these problems stemmed from rural discontent, and rural discontent stemmed from hunger" (Fowler/Mooney 1990: 56), for which there could be no military solutions.

In the eyes of Western agronomists, the small-scale and subsistence-oriented farming system predominating in these regions was both economically inefficient and unable to implement improvements they considered necessary to quickly increase agricultural production. Rather than supporting farmers and land reforms, therefore, the Green Revolution development strategy pursued a top-down approach that consisted in strengthening governments and national agricultural institutions as well as in bringing in agronomic "experts" from the industrialised world. At the same time, efforts were made to transform "potentially volatile peasants into cooperative capitalist farmers" (Curry 2017b: 2). This was achieved by, among other things, discursively constructing "far-away communists capitalizing on the poverty of the countryside [...] as the main enemies of global progress towards peace and prosperity" (Flitner/Heins 2002: 328). This shows that the ideas about agricultural development that scientists and agronomic experts from 'developed countries' imported into the 'developing world' through the Green Revolution are deeply rooted "in contemporary theories of economic development and [...] bound closely to US geopolitical concerns and strategies for national security in the midst of the Cold

War" (Curry 2017b: 2). The breeding and introduction of 'improved' high-yielding crops in the 'developing world', then, not only served to fight hunger but was a developmentalist bio- and geopolitical instrument to establish Western and capitalist hegemony after World War II (Flitner/Heins 2002: 320).

Although agricultural and food production indeed increased in the context of the Green Revolution, according to Fowler and Mooney, hunger persisted and even increased in many of the areas where the new high-yielding varieties were grown. They explain this discrepancy as follows:

> "From the beginning they did not grow as well for the poor farmers as for the richer ones. Achieving high yields required fertilizer and irrigation. Fertilizer and irrigation nourished weeds as well as crops, creating the need for herbicides. And pests found the uniformity of new varieties appetizing, which necessitated the use of insecticides as well. Farmers lacking access to capital to buy these items were simply left in the dust." (Fowler/Mooney 1990: 58)

For Fowler and Mooney, it follows that one of the problems with the Green Revolution was that "it offered a technological solution to a social and political problem" (ibid.). What is more, this solution was developed in a world that essentially differed from the worlds in which it was implemented. In consequence, "[f]or many a Third World farmer, it was not just a new way of farming that had to be learned, but a whole new world view that had to be assimilated" (Fowler/Mooney 1990: 75). Instead of supporting existing self-provisioning food systems by integrating local farming systems into the production process and providing produce to the hungry through effective distribution, the developmentalism of the Green Revolution created dependencies that ultimately strengthened the industrial sector in the 'developed' world (see Fowler/Mooney 1990: 130).

It becomes discernible that the Green Revolution, as Marci Baranski argues, "was not a linear transfer of technology but rather a result of coinciding sociopolitical forces including the Cold War, technological optimism, the rising status of plant breeders, new institutions in international aid and development, and changing agricultural policies" (2022: 11). Or, as Fowler and Mooney put it, it was not "something as clean as the 'modernization of agriculture' or as simple as the importation of new varieties" (1990: 76). Following these critical accounts, I understand the Green Revolution as a performative manifestation of a particular approach to doing agriculture. This approach was historically and socio-politically situated in the technologically and science-driven post-war "development project" (McMichael 2019) that the Western world began to cultivate in the mid-twentieth century to consolidate the global capitalist economy. The euphemistic promises of 'improvement' and 'revolution' discursively obfuscate the imperialist underpinnings of a developmentalism purporting to be in the interest of humanity at large. They obscure that what was homogenised through the breeding and large-scale introduction of 'improved' varieties were not only seeds and agricultural products but also the agri-cultures practiced in different parts of the world.

In the world of crop plants, the homogenisation and interspersion of agricultural products and production with chemicals resulted in the destruction of habitats as well as in the replacement of traditional landraces with "modern 'cash crop' varieties" (Fowler/Mooney 1990: 76). According to Fowler and Mooney, the devastations and the threat to crop diversity that come with this development are hard to overestimate:

> "The genetic diversity being lost today is the foundation of future plant breeding, of future plant evolution. If enough diversity is lost, the ability of crops to adapt and evolve will have been destroyed. We will not have to wait for the last wheat plant to shrivel up and die before wheat can be considered extinct. It will become extinct when it loses the ability to evolve, and when neither its genetic defenses nor our chemicals are able to protect it. And this day might come quietly even as millions of acres of wheat blanket the earth." (Fowler/Mooney 1990: 89)

The following section discusses the social-ecological crises provoked by this progressively advancing extinction – or erosion, as it was eventually termed – as well as the rise of seed banking as a countervailing measure embodying the promise of stabilising modern agriculture.

1.2 Genebanks Against Genetic Erosion: The Proliferation of *ex situ* Conservation

With the FAO, the first intergovernmental organisation dedicated to globally coordinated agricultural and food policy entered the political arena in 1945. In its early years, the FAO's efforts were concentrated on plant exploration, breeding, and the introduction of new varieties. The FAO conference of 1959 was the first to acknowledge the potential ecological threat arising from the homogenisation of crop cultivation and the ensuing de-cultivation of diversity in the field (Flitner 1995: 150).[8] As a consequence, a Panel of Experts in Plant Exploration and Introduction (hereafter: Panel of Experts) mandated to advise the FAO on the matter was created in 1965.[9] These experts, who represented different scientific and agro-political institutions and organisations, met regularly over the following decade to develop international guidelines for the collection, conservation and exchange of seeds (Esquinas-Alcázar/Hilmi/López Noriega 2012: 136). Based on the work of the Panel of Experts, in 1967 the FAO co-organised a Technical Conference on the Exploitation, Utilization and Conservation of Plant Genetic Resources together with the International Biological Programme (IBP).[10] It was during this conference that the increasing loss of genetic diversity in agricultural plants was first recognised and defined as a global ecological problem under the newly coined term "genetic erosion" (Frankel et al. [1969]

---

8 The biannually convened FAO conference is "the highest decision-making body in the organisation, where all member countries are represented" (Esquinas-Alcázar/Hilmi/López Noriega 2012: 136).

9 The members of this panel of experts were Otto H. Frankel, former director of the Australian Commonwealth Scientific and Industrial Research Organization and head of the working group on plant genetic resources of the International Biological Programme (IBP); Erna Bennett, a representative of the FAO and director of its Crop Ecology and Genetic Resources Unit from 1967 to 1982; Jack Harlan, a plant geneticist from the University of Illinois; Jack Hawkes, a potato-specialist from the University of Birmingham; John L. Creech from the New Crops Research Branch in Beltsville; and Hermann Kuckuck, plant geneticist and Professor of Applied Genetics at the University of Hannover until 1969 (see Flitner 1995: 152; Peres 2016: 100).

10 The IBP was a research initiative started in 1964 on behalf of the International Council of Scientific Unions and UNESCO. It countered the FAO's resource-oriented approach to plant genetic resources with an information technology-informed ecosystemic perspective. According to Fenzi and Bonneuil, "[t]he IBP promised expertise toward the optimal tuning of natural productivity cycles of the biosphere, to achieve a static state accommodating short-term economic demand, as well as maintaining processes that renew natural resources in the long term" (2016: 76). However, further research in the wake of the 1967 FAO/IBP Technical Conference identified the dissemination of modern cultivated varieties as the main cause of genetic erosion, which strengthened plant genetic resources conservation rather than crop ecology, as I elaborate below (see also ibid.). Joanna Radin (2013) intriguingly discusses the biopolitics of the IBP, focusing on a bio-anthropological and genetical subproject exploring human adaptability to changing conditions of life.

1970; Frankel/Bennett 1970).[11] While it poignantly captures the extent of the problem, the notion of genetic erosion also implies a passivity that obfuscates the ways in which it arises from specific agricultural modernisation practices (see also Fowler/Mooney 1990: 54–89; Fenzi/Bonneuil 2016: 74–76).

Another term that took root at this time was the notion of "genetic resources", which became the primary reference for the diversity of plant material targeted in conservation activities. Historians have pointed out that the term encapsulates the (molecular-)genetic focus predominant in plant science since Vavilovian times (Fenzi/Bonneuil 2016: 73–74; Peres 2016: 101–102; see also Alpsancar 2016: 52).[12] In addition, it explicitly denotes the economically oriented system of agriculture and its resourcist approach to nature that fuels the problem of genetic erosion, that is to say, the loss or destruction of plant (genetic) diversity (Flitner 1995: 152–153). What is more, according to Michael Flitner, the terminology further indicates an emerging consciousness of humanity's dependence on finite natural resources and global interdependence: "The view of plant diversity as a *resource*, which, like soil, is threatened by *erosion*, was a harbinger of the discourse on the 'limits to growth' that emerged not long after this, and fits into conceptions of the biosphere emerging at that time" (Flitner 1995: 154, author's translation, emph. in orig.).

With the proliferation of the term genetic resources, Flitner (1995: 155) argues, the FAO also dropped the discursive distinction between plant exploration and plant introduction, thus shifting the focus away from the divergent origin and flow of genetic resources. As a result of the growing awareness of globally advancing genetic erosion, collections of plant genetic diversity went from serving primarily as resources for breeding new improved varieties to becoming crucial components for managing the undesired consequences of the Green Revolution (Curry 2017b: 1–3). The FAO/IBP conference of 1967 sought to replace the purpose-oriented approach to collecting and conserving plant genetic diversity as a stock of resources for breeding with a generalist and long-term conservation approach so as to secure the largest possible diversity of en-

11 According to Flitner (1995: 152–153), it can be assumed that the term was coined by the Panel of Experts, who prepared and evaluated the conference and wrote the report that makes extensive use of it.

12 Bonneuil (2019) unfolds "a deeper history of 'genetic resources' as a category, i.e., the emergence of a genocentric and resource-centric ontology of the planet's biological diversity" (2019: 1).

dangered landraces.[13] In the wake of this conference a "genetic resources movement" took shape (Kloppenburg [1988] 2004: 163; Pistorius 1997). Over the following years, there were heated debates within this movement about the best approach to plant genetic diversity conservation.

Proponents of *in situ* conservation strategies campaigned for conserving diversity through cultivation in the field, arguing that plants evolve in and through a complex and dynamic relationship with their (constantly changing) environments. In the Panel of Experts, *in situ* conservation was strongly defended only by FAO representative Erna Bennett.[14] Most other panellists, with Otto H. Frankel leading the way, were sceptical about *in situ* conservation strategies on financial, technical, and social grounds. Frankel expressed doubt about whether farmers in the Global South would continue to cultivate landraces despite the availability of higher-yielding varieties, about the suitability of the *in situ* approach for long-term conservation efforts that would be able to halt and prevent further genetic erosion, and about the compatibility of "*in situ* conservation [...] with the contemporary modernisation of agriculture" (Peres 2016: 100).

Ex situ conservation in seed banks, in contrast, received broad support in the FAO and the genetic resources movement more generally for at least two reasons: first, the expectation that this conservation strategy could make available for the long term a diversity of plant genetic resources that was economically unappealing for cultivation in the present; and second, the widely shared conviction that the high productivity and efficiency of industrial and monocultural production made the continuation of modern agriculture non-negotiable.[15] *Ex situ* conservation in seed banks was therefore adopted by the FAO as

---

13 Flitner (1995: 155–156) notes that while the guiding principles and strategies developed at the 1967 FAO/IBP conference did inform conservation activities through the 1970s and 80s, in very few cases were seed collections actually generalist rather than mission-oriented.

14 Despite the difficulties of making a career as a woman at that time, Bennett was a key figure in the plant genetic resources movement. After attracting attention with two influential publications that introduced the "genecological" position and the concept of "genetic conservation" in 1964 and 1965, she joined the FAO in 1967, prepared the FAO/IBP technical conference that same year, and co-edited the proceedings together with Otto H. Frankel (Frankel/Bennett 1970). In 1982, she was forced to leave the FAO due to political differences, and went on to comment on later developments from outside UN bodies (Hanelt/Knüpffer/Hammer 2012; see also Pistorius 1997).

15 Pistorius (1997: 24–27) further explains that the two approaches are rooted in diverging views of how plants adapt to changing environmental conditions. While *in situ* conser-

the core strategy to save and preserve plant genetic diversity (Fenzi/Bonneuil 2016: 75; Flitner 1995: 152–157).

According to historians Marianna Fenzi and Christophe Bonneuil (2016: 77), the preference for *ex situ* conservation is indicative of an agricultural regime defined by theories of modernisation, in the context of which breeding is a logical, effective, and pragmatic development of agriculture. Volker Heins and Michael Flitner (1998: 21–22) attribute the strong support for the *ex situ* approach in the FAO to three main factors. First, gene-oriented breeding was economically interesting because it allowed for creating varieties with short production cycles. Second, political upheavals in the Global South evoked fears in the Global North that the end of the current extractive colonial botany was imminent, and that this would complicate access to the abundance of plant genetic diversity mostly prevalent in the 'developing countries'. Third, the development of technologies of cold storage had advanced in a way that made long-term independence from field cultivation seem possible.

To realise the vision of a global-scale and long-term *ex situ* conservation solution for the world's endangered plant genetic diversity, the 1967 FAO/IBP Technical Conference devised a recommendation to create or designate an international seed storage facility available to every country in the world (Curry 2017b: 5; Flitner 1995: 156). However, the capacity of the FAO and the IBP to put this recommendation into action was hindered by a lack of funding and cooperation on the part of other institutions engaged in international agricultural policy and development (Curry 2017b: 12). As a consequence, the FAO incrementally lost its leadership role in the field of international conservation efforts throughout the 1960s and 70s.

Instead, financially strong actors such as the Rockefeller and Ford foundations, which were much more economically oriented than the FAO, gained a foothold. They preferred to work with the international plant research and breeding centres (the IARCs) that had been established as part of the Green Revolution and strengthen their approaches (Flitner/Heins 2002: 328). As a result, the reality of conservation activities became much narrower than envisioned at the FAO/IBP conference insofar as they became oriented towards the interests of breeders much more than towards "advancing food security through the conservation of a broad range of crops and the distribution of this

---

vation is based on a "genecological" approach that focuses on the interaction of polygenes with their environment, *ex situ* conservation supports controlled adaptation of single genes to changing environments through breeding.

work and the resources necessary to support it to a range of national and international actors" (Curry 2017b: 16). Importantly, these institutions did not consider breeders as "stakeholders with their own interests" so much as "operating for the general interest, transforming a common heritage of resources into collective food output" (Fenzi/Bonneuil 2016: 75). In line with this, their agronomists considered replacing landraces with improved varieties as a logical component of the modernisation of agriculture and therefore saw genetic erosion as "an inevitable part of economic and social development" (ibid.).

Around the same time, the public – beginning in the Global South and soon also in the Global North – got a bitter taste of the fatal consequences of the economic and social 'developments' that followed in the wake of the Green Revolution. With the homogenised cultivation of the newly bred improved crop varieties came a uniformity in crops that increased genetic, ecological, and economical vulnerability as well as the severity of the crises that would eventually result from them. By the 1970s, almost half of wheat production in the Global North relied on wheat varieties bred at the Mexican International Maize and Wheat Improvement Center (CIMMYT) (Fowler/Mooney 1990: 69). A game changer in agricultural practice and politics was the global spread of the fungal plant pathogen *cochliobulus heterostrophus*, which causes southern corn leaf blight and ruined maize cultivation around the globe beginning in the 1960s. In 1970, it arrived in the US, where it destroyed fifteen per cent of the corn yield, over ninety per cent of which consisted of varieties particularly vulnerable to this pathogen, and caused enormous costs for both farmers and consumers (Fowler/Mooney 1990: ix; Kloppenburg [1988] 2004: 163).[16]

Ironically, Norman Borlaug received the Nobel Peace Prize for his breeding of high-yielding semi-dwarf wheat varieties at the Mexican CIMMYT the same year and the World Bank suggested the creation of a broader financial network for the IARCs. The realisation of this proposal resulted in the creation of the Consultative Group on International Agricultural Research (CGIAR) in 1971, which connected the four IARCs hitherto established in Mexico, Columbia, Nigeria, and the Philippines as well as two newly founded IARCs in Peru and

16 A study by the US National Academy of Sciences published two years later and titled "Genetic Vulnerability of Major Crops" called attention to the genetic uniformity and resulting high vulnerability of US agriculture, which depended on only a handful of genetic varieties of their major crops (see Fowler/Mooney 1990: 83).

the Ivory Coast under its roof.[17] The most important difference between the CGIAR and the FAO as a UN organisation was and is that political control of the latter lies with all member states whereas the CGIAR is politically controlled by its donor countries – most of which are located in the Global North (Flitner 1995, 168). In 1972, the CGIAR convened a conference in Beltsville, Maryland (USA), to discuss the recent devastations caused by the southern corn leaf blight. Erna Bennett, who represented the FAO on the Panel of Experts and was the key proponent of *in situ* conservation in the genetic resources movement, was not invited. In lieu of the creation of one central world genebank, as previously suggested by the FAO, the Beltsville conference recommended the creation of a network of nine major genebanks. However, the technical committee of the CGIAR rejected this proposal and instead opted for the creation of a subsidiary network to the CGIAR to take on the responsibility for developing countermeasures against genetic erosion on a global scale (Flitner 1995: 169–171).

In 1974, the CGIAR created an International Board of Plant Genetic Resources (IBPGR), which was tasked with coordinating a rescue programme for plant genetic resources (PGR) based on the CGIAR's policy on genetic resources presented in Beltsville. This policy focused on exploring potentially agronomically useful wild and cultivated plant species and conserving them *ex situ* for future use. Financial support, personnel, and premises were supposed to be provided largely by the FAO in the form of its Crop Ecology Unit. While the United Nations Environment Programme (UNEP)[18] was included in the supervisory board, control over the IBPGR remained mostly in the hands of the CGIAR with its insufficient democratic structure (Flitner 1995: 172). Based on expert knowledge and recommendations, the IBPGR drew up a list that prioritised crops for conservation according to their level of endangerment, representation in existing collections, and social and economic importance, as well as breeders' needs (see Fowler/Mooney 1990: 79–80). Notwithstanding the fact that this list was certainly biased towards the needs of industrialised countries, Fowler and

17 Today, there are 15 CGIAR centres in total. For more information see https://www.cgiar.org/research/research-centers/ (last accessed July 18, 2025).

18 The UNEP was founded during the first UN Conference on the Human Environment (UNCHE) in Stockholm, Sweden, in 1972. It was supposed to become the responsible actor for genetic resources among the UN organisations. However, the CGIAR had already pushed the structural decision-making process to a point that left the UNEP with little to no influence (Flitner 1995, 171).

Mooney (1990: 79) concede that it was based on the best educated guesses at the time.

With the formation and collecting efforts of the IBPGR, the by then twelve research centres able to store genetic resources and make them available to plant breeders, and the vast financial resources flowing into the CGIAR, the latter superseded the FAO as leader of the global management of plant genetic resources (Fenzi/Bonneuil 2016: 77). Civil society actors criticised the creation and advent to power of the CGIAR as "a blunt move by these donors to wrest control of agricultural development from FAO and place it in the hands of a manageable scientific élite" (Mooney 1983: 66, quoted by Flitner 1995: 168). Flitner (1995: 167–169) comments critically that while this judgement may have been politically apposite, the lack of financial resources and determination within the FAO to assume the role that was then delegated to the IBPGR made it impossible for the FAO to arrive at a comparable result. What is more, he questions whether the agrarian developmental politics the FAO represented were really that different from the CGIAR's, in view of the fact that the corporate sector gained increasing influence in the FAO from the mid-1960s onwards (Flitner 1995: 172–173).

1.3 Conserving the "Heritage of Mankind"? Politicising the Loss and Conservation of Plant Genetic Resources

As the previous section has shown, the issue of genetic erosion was mainly discussed on a technical level throughout the 1960s and 70s. This was also indicated by the labelling of the FAO conferences in 1967, 1973 and 1981 as "technical conferences". Beginning with the FAO conference in 1979 and through the 1980s, discussions about the internationally concerted efforts to conserve endangered plant genetic diversity – including the discussion about establishing an international seed bank or seed bank network – became more politicised. During this time, non-governmental and civil society organisations entered the agro-political arenas in the UN and CGIAR environment, politicising what had up to then been intergovernmental and public-private debates about agrobiodiversity conservation in new ways. These organisations supported peasant farmers, mainly but not exclusively in the Global South, in fighting against technocratic and economistic developments that exacerbated global social inequalities and ecological problems.

At the forefront of these organisations was the Rural Advancement Foundation International (RAFI),[19] the first civil society organisation (CSO) "to draw attention to the socioeconomic and scientific issues related to the conservation and use of plant genetic resources, intellectual property and biotechnology" (ETC Group n.d.; see also Fenzi/Bonneuil 2016: 77). RAFI was founded by Fowler and Mooney, together with Hope Shand, in the 1970s with a focus on intellectual property and biotechnology issues in agrobiodiversity conservation.[20] In 1979, Mooney published a pathbreaking book titled *Seeds of the Earth: A Private or Public Resource?* which criticised the exploitative politics around plant genetic resources for food and agriculture (PGRFA) pursued by some of the 'developed countries', as well as the consolidation of transnational agro-industrial corporations into a growing agribusiness sector. As Flitner and Heins summarise, Mooney argued that these developments advanced the formation of two new major oppositional geopolitical blocs: "the 'North' represented by transnational corporations from the capitalist world, and the 'South' exemplified by traditional small-scale farmers being deprived of their livelihood resources" (Flitner/Heins 2002: 329–330). Mooney's arguments were influential for the ensuing controversies within the FAO and, in the long term, for the support of multilateral solutions that eventually emerged (see also Andersen 2008: 90; Fenzi/Bonneuil 2016: 77).

In the early 1980s, however, a North-South-polarisation took shape within the FAO that would last for years and have a considerable influence on how later international agreements on PGR conservation were negotiated. At the FAO conference in 1979, the Spanish delegation tabled the very first concrete proposal in the history of agrobiodiversity conservation for a centralised international genebank under the aegis of the FAO and an international agreement on plant genetic resources. The proposal was not turned into a draft resolution, though, due to concerns voiced by a coalition of 'developing countries', which shaped the subsequent discussions in the international agro-political arena. Based on the fact that "[p]lant genetic resources are found throughout the world, but the greatest diversity is in tropical and subtropical areas, where

19 RAFI was renamed Action Group on Erosion, Technology and Concentration (short: ETC Group) in 2001.

20 A well-known member of the founding board was Erna Bennett, who was still working for the FAO at that time. In the 1990s, the focus of RAFI's research, education, and social action extended to environmental concerns, biopiracy, and genetics/genomics as well as later to nanotechnology.

most developing countries are situated" (Esquinas-Alcázar/Hilmi/López Noriega 2012: 137), these countries were concerned about two issues in particular. For one thing, they questioned the legal circumstances established by a convention of the International Union for the Protection of New Varieties of Plants (UPOV) adopted in 1961, which granted intellectual property rights for crop breeds to breeders and protected these rights while not legally recognising the contribution of farmers cultivating the raw material for the breeding of new varieties. Secondly and as a consequence of this legal situation, there were concerns about whether resources collected in an international seed bank or seed bank network would be considered property of the country of origin, the country in which they would be stored, or humanity at large (Fenzi/Bonneuil 2016: 76).

As a result of these concerns about the Spanish proposal, in 1981 the Group of 77, a coalition of 'developing countries' within the FAO, presented a draft resolution conceptualised in cooperation with the state of Mexico and the Group of Latin America and the Caribbean (GRULAC) aiming to establish an international germplasm bank in Mexico.[21] After a heated debate about this proposal that lasted for several days, a working group was created and tasked with conducting feasibility studies and developing a revised, more consensual resolution. According to these studies, an international genebank or a genebank network were not technically feasible and a legally binding agreement not necessary, whereupon this second proposal was rejected at the FAO conference in 1983 (Esquinas-Alcázar/Hilmi/López Noriega 2012: 136–138). José Esquinas-Alcázar and colleagues argue that pinning the problem on technical feasibility actually served to obfuscate the lack of political will to follow up on the draft resolution, which became apparent in yet another proposal made by the Spanish delegation at the end of these discussions, namely to place their national genebank under FAO jurisdiction (see Esquinas-Alcázar/Hilmi/López Noriega 2012: 137).[22] Although this proposal, too, was rejected, the discussions within

21 The term "germplasm" refers to genetic resources maintained for breeding. While germplasm is contained in seeds, it is not equivalent to them. Therefore, facilities for *ex situ* conservation, research, and breeding, such as the IARCs under the umbrella of the CGIAR, are referred to as "genebanks" rather than "seed banks".

22 José "Pepe" Esquinas-Alcázar first started working for the FAO in 1978 and became secretary of the FAO's intergovernmental Commission on Plant Genetic Resources (CPGR) in 1983, a position that he occupied until his retirement in 2007. He further chaired the FAO Committee on Ethics for Food and Agriculture between 1999 and 2007, and was the first Secretary of the FAO's International Treaty on Plant Genetic Resources for Food and

the FAO had sown the seeds "for an international convention on plant genetic resources as well as a new 'international gene bank'" (Fowler/Mooney 1990: 187). A resolution to this effect was passed at the next FAO conference in 1981.

One of the driving forces behind the efforts to internationalise plant genetic resources conservation under FAO jurisdiction in a way that protected the interests of the Global South was a widespread critical attitude towards the IBPGR. Within the first ten years of its existence, the IBPGR had conducted 500 expeditions and collected 100,000 varieties, 90 per cent of which originated in the Global South, whereas only 15 per cent of these varieties were stored in the genebanks of the respective countries (Flitner 1995: 191).[23] Many therefore perceived it as an instrument used by the industrialised countries of the Global North to access the resources of the Global South, commodify them as privatised varieties, and thus take freely accessible 'raw material' off the market. It was in this context that the resource-rich countries of the Global South claimed national sovereignty and control over their plant genetic resources in an effort to strengthen the "centres of origin" rather than the "centres of agricultural modernization" (Flitner/Heins 2002: 327). However, without an international regulatory framework for the conservation, utilisation, and exchange of plant genetic resources, the discussion about sovereignty and property rights remained without effect.

By the end of 1983, an agreement was reached among the vast majority of the FAO's member states that gave rise to the first International Undertaking on Plant Genetic Resources (IUPGR; hereafter: IU for International Undertaking). This legally non-binding agreement, signed by over 100 of the FAO's member countries with the exception, among others, of the USA and Canada,

---

Agriculture (ITPGRFA, short: Seed Treaty, see below) between 2004 and 2007. One of my interview partners referred to him as "the Don Quijote fighting all these mysterious windmills of conservation" (Alejandro Argumedo, Potato Park, 5 November 2020: l. 530–532).

23 The criticism of this practice ties in with a critical discourse on "biopiracy", that is, "the fraudulent acquisition of ownership over genetic materials" (Breen 2015: 42), which reaches back to the "Columbian Exchange" (Crosby 1972) and the colonial history of botanical gardens (Brockway 2002; Philip 2004). Biopiracy, according to Jack R. Kloppenburg (2000: 511), is one of "the contemporary naming[s] of the current manifestation of an activity with deep historical roots. Crop genetic resources and medicinal plants have long been freely taken from peasant farmers and indigenous peoples either forcibly or, more recently, with the justification that such materials are the 'common heritage of mankind'."

was supposed to advance international cooperation in agricultural politics and promote the exploration, conservation, evaluation and availability of plant genetic resources for scientific research and plant breeding. An intergovernmental Commission on Plant Genetic Resources (CPGR), referred to as "the Commission" within the biodiversity conservation community, was created to monitor and later renegotiate the IU (Esquinas-Alcázar/Hilmi/López Noriega 2012: 137–138; Flitner 1995: 191–193; Fowler/Mooney 1990: 188).[24] Defining the objective of the IU, article 1 of the approved text established "the universally accepted principle that plant genetic resources are a *heritage of mankind* and consequently should be available without restriction" (FAO 1983, emph. added). Article 2 states that this encompasses cultivated varieties, obsolete cultivars (i.e. varieties no longer cultivated), landraces, and wild relatives as well as "special genetic stocks (including elite and current breeders' lines and mutants)" (FAO 1983), all of which should be accessible without let or hindrance, according to article 5 (see also Esquinas-Alcázar/Hilmi/López Noriega 2012: 138).

While Fowler and Mooney celebrate the IU for establishing the first-ever intergovernmental body "capable of applying not only scientific but also social and political realities to the collection and exchange of genetic resources" (1990: 188) thanks to which "[t]he world's donors of germplasm finally have a voice" (ibid.), others expressed reservations. Some representatives of the industrialised countries and private seed industry of the Global North fiercely resisted the provisions of the IU, which they saw as destabilising plant breeders' rights and the protection of improved varieties as regulated by the UPOV convention (Flitner 1995: 194). Some representatives of the resource-rich countries of the Global South also had reservations about the IU and the CPGR, albeit for a different reason: the denomination of plant genetic resources as a "heritage of mankind" subverted the idea of national sovereignty they had long been fighting for (Flitner 1995: 194–195).[25]

---

24 The Commission was renamed as the Commission on Genetic Resources for Food and Agriculture (CGRFA) in 1995. Both the Commission and the IU were established without consensus and against the will of eight FAO member states from the Global North (Esquinas-Alcázar/Hilmi/López Noriega 2012: 137–138, 146n4; Flitner 1995: 192–193).

25 Flitner (1995: 195–196) points out that the attribute "common" in the more familiar formulation "common heritage of mankind" wasn't added until the 1989 resolution on farmers' rights. According to him, this can be traced back to doubts voiced on both sides of the controversy. He quotes Kloppenburg and Kleinman (who overlook the fact that the attribute "common" was not part of the original formulation in the IU) to point

Notwithstanding these reservations, the IU and the CPGR furthered international cooperation in the exploration, preservation, evaluation, and availability of plant genetic resources for plant breeding and scientific purposes. In the years that followed, the CPGR negotiated three resolutions that were annexed to the IU, all of which were major achievements from the viewpoint of the Global South. In 1989, the IU was amended by two annexes that recognised the rights of farmers as equal to those of plant breeders. While an instrument to protect breeders' rights had been established many years ago in the form of the UPOV convention, the revised IU constituted the first official international acknowledgment of "the enormous contribution that farmers of all regions have made to the conservation and development of plant genetic resources, which constitute the basis of plant production throughout the world" (FAO 1989).[26] In order to follow up this recognition with actions, the CPGR agreed to create an international fund to support farmers, the extent of which should be specified in consequence of a quantification of the receiving countries' needs. A third resolution, adopted in 1991, revisited "the concept of mankind's heritage, as applied in the International Undertaking on Plant Genetic Resources, [defining it as] subject to the sovereignty of the states over their plant genetic resources" (FAO 1991). This dispelled the abovementioned concerns about the IU in the Global South, and put the question of restrictions of access to these resources back on the table (see also Andersen 2003: 47–48).

In the late 1980s William Brown, chair of the US National Board for Plant Genetic Resources and an influential figure within the seed industry, invited the various rival agro-political parties to the Keystone Center in Colorado,

---

out the irony of this controversy: "[I]n a world economic system based on private property, each side in the debate wants to define the other side's possession as common heritage. The advanced industrial nations of the North wish to retain free access to the developing world's storehouse of genetic diversity, while the South would like to have the proprietary varieties of the North's seed industry declared a similarly public good." (1988: 188; see also Flitner 1995: 196)

26 An ambiguity of this new acknowledgment of farmers' rights, as Flitner (1995: 198) briefly discusses, lies in the insufficient practical implementation of these resolutions. While they recognised and promised to support the work of peasant farmers, it remained unclear whether financial compensations paid to national institutions were cashed out to the farmers themselves, especially since the governments of many 'developing countries' were at the time pursuing a policy of agricultural modernisation incompatible with this international agreement. According to Flitner, as an intergovernmental arena, the FAO was ultimately unable to operate beyond the scopes of nation-states.

USA, a mediation centre for conflict resolution in contentious political matters founded in 1975. In the Keystone Dialogues, it brought international stakeholders together as individuals rather than representatives of their respective organisations for off-the-record discussions of controversial issues aimed at producing a consensual report. In three meetings between 1988 and 1991, the international agro-political community convened in this face-to-face and consensus-oriented negotiation space with a view to pacifying the heated controversies (Andersen 2008: 95; Fowler 1994: 196–204). According to Fowler, while questions around intellectual property rights remained largely unresolved, "important ground was broken in two areas – in defining the notion of genetic resources as 'common heritage' and in the emerging concept of 'farmers rights'" (1994: 199). This influenced the FAO resolutions developed during that time to amend the IU. Apart from concrete results, according to Fowler (1994: 200), what the Keystone Dialogues also accomplished was a more constructive and mutually respectful atmosphere among the representatives of the various interest groups from the seed industry to the non-governmental sector, which was even more important for future negotiations.

In the controversial climate characterising PGR politics in the 1980s and early 1990s, another initiative to establish an international PGR conservation facility took shape, this time in the Arctic. This is rarely mentioned in the history of internationally coordinated conservation efforts, presumably because, like previous attempts, it was unsuccessful. However, it is an important step to consider when tracing the history of the global efforts in PGR conservation. For one thing, it illustrates the entanglement of technical and political matters in biodiversity conservation, and for another, it laid the groundwork for the eventually successful establishment of the Svalbard Global Seed Vault in 2008. In 1988 and 1989, the IBPGR first investigated the possibility of establishing a facility in the Arctic permafrost for storing safety duplicates of the IARCs' seed collections. The idea was inspired by the Nordic Gene Bank for Agricultural and Horticultural Plants, which had been using an inoperative coalmine on the Arctic island of Spitsbergen in the Svalbard archipelago for secure storage of a backup of their collection since 1984.[27]

27 The Nordic countries include the states of Denmark, Finland, Iceland, Norway and Sweden, as well as the Faroe Islands and Greenland (both part of Denmark), the Åland Islands (an autonomous archipelago that belongs to Finland), and the Svalbard archipelago, which belongs to the sovereign territory of Norway. The Nordic countries have a long history of political and other kinds of collaboration, including cooperation

It was first and foremost the cost-effectiveness of storing seed collections inside a permafrost mountain, guaranteeing year-round sub-zero temperatures without the need for technical cooling, that aroused the IBPGR's interest in the region (IBPGR 1991: 10).[28] After conferring with experts from Norway about legal and technical feasibility, the IBPGR decided to pursue the project of establishing a security storage facility under the name Svalbard International Seedbank (SIS). After consultations between the IBPGR and the CPGR, the FAO joined the project in 1989.[29] A questionnaire was sent out to 750 international experts and institutes to determine the extent of interest in such a facility. While the IBPGR's annual report of 1990 states that "[p]ositive responses came from half the respondents [and] [a]pproval of the other half was mostly conditional on the clarification of minor issues" (IBPGR 1991: 10), Marte Qvenild

---

in agricultural issues. In 2007, the Nordic Genebank for Agricultural and Horticultural Plants merged with the Nordic Genebank Farm Animals and the Nordic Council for Forest Reproductive Material and became the Nordic Genetic Resource Center (NordGen), one of the three partners responsible for the Svalbard Global Seed Vault (see NordGen n.d. a).

28 A delineation of the proposal in the IBPGR's annual report of 1990 specifies an average temperature of -3.7 degrees Celsius inside Svalbard's permafrost mountains (IBPGR 1991: 10). Information on the average temperature in Svalbard and of the permafrost varies from source to source. This, at least in part, goes back to the fact that the Arctic is subject to increasingly severe climate changes (see e.g. Øseth 2011; AMAP 2012; Norwegian Polar Institute n.d.).

29 In the few publications about the project that go beyond merely mentioning it, the SIS initiative is depicted as a joint effort of the IBPGR and FAO (Qvenild 2008) or of the FAO and other "international experts" (Qvenild 2012: 304). While it was a joint project, it seems worth mentioning that the initiative came from the IBPGR and not the FAO. Qvenild (2005) elaborates this in her Master thesis, submitted to the Department of International Environment and Development Studies of the Norwegian University of Life Sciences, for which she conducted the study underlying the abovementioned articles. Qvenild offers a detailed description of the Nordic and international efforts to store seed collections in the Arctic permafrost, based on interviews with a range of experts from the field of plant genetic resources conservation including people involved in the first Svalbard initiative. She quotes a statement from the director of the IBPGR at the time, made in a report on the expert consultation on SIS held at the IBPGR headquarters in December 1990: "[...] IBPGR did not have the right intergovernmental status to conclude a worldwide agreement on security storage in permafrost and I approached FAO in the second half of 1989 to assess FAO's interest in joining in this effort. This proposal was presented in 1989 to the Working Group of the FAO Commission, which, in principle, accepted the proposal. From that point onwards, this has been a truly joint IBPGR/FAO effort." (Qvenild 2005: 38)

(2008: 113) quotes an unpublished FAO/IBPGR Report of the Expert Consultation on SIS from 1990 indicating a response rate of no more than 19 per cent, including a number of negative replies that questioned the relevance of such a facility (for a more detailed account see Qvenild 2005: 39). Despite the low response rate, the IBPGR and FAO pursued the project, prioritising technical over political feasibility issues. With the Norwegian government's assured financial support for the construction of the facility, the FAO and IBPGR agreed to cover the costs of operations and administration and to create an Endowment Fund of 15 million dollars to secure funding for the first 50 years (Qvenild 2008: 113).

The SIS project was ultimately abandoned. According to Qvenild (2005, 2008), financial and technical constraints prevented the realisation of the project, but political concerns were the main reasons that prompted potential donors to back out. During a time of polarised political discussions about access to PGR and property rights, "plac[ing] an 'international' storage bank in a developed country without legal clarity about the ownership of the material meant, in the eyes of NGOs and developing countries, giving the material away" (Qvenild 2008: 112). The proposal attempted to settle these doubts by promising that all deposits made to the SIS would remain property of the depositor and non-accessible to others without consent, while remaining always and unrestrictedly accessible to the depositor. In the absence of legally binding international regulations, however, the plant genetic material stored in the SIS ultimately would be subject to Norwegian legislation. Against the background of the historical experience that most plant genetic material collected in the Global South by the IBPGR and its precedent institutions had been placed in genebanks in or controlled by the Global North and used by the Western seed industry for the breeding of improved varieties protected through intellectual property rights without sharing the profits or granting free access to these varieties, the resource-rich countries from the Global South in particular distrusted the promises made about the SIS. This atmosphere of suspicion negatively impacted the commitment of previously supportive potential donors and increased already existing scepticism about a crucial technical issue: although the IBPGR's own *Genebank Standards* (1984) defined -18 degrees Celsius as the optimal temperature for long-term storage, there was at that time no technical concept that would have made it possible to cool the seeds down to a lower temperature than the -3.7 degrees Celsius of the permafrost.

The project was finally dropped in 1994 (Qvenild 2005: 47) after another obstacle had emerged with the Convention on Biological Diversity (CBD) in 1992. Although the CBD contributed to the consolidation of international agro-political cooperation in the medium term, as the following section outlines, in the first instance it created a restrictive and bureaucratically complex environment for the regulation of access to and exchange of PGRFA (Qvenild 2008: 114).

1.4 From the Convention on Biological Diversity to the International Seed Treaty: Consolidating International Cooperation

After the FAO had acknowledged the placement of plant genetic resources under national sovereignty through the revision of the IU in 1991, the UN ratified this development one year later in the CBD, the first binding international agreement on biological diversity (see Secretariat of the CBD [1992] 2011). A Working Group of Experts on Biological Diversity within the UNEP, established in 1988, had spent four years negotiating this agreement. The CBD was submitted for signature at the UN Conference on Environment and Development (UNCED) – the first Earth Summit – in Rio de Janeiro, Brazil, in June 1992 (Fenzi/Bonneuil 2016: 77). While the CBD is one of the most important developments in international politics around biodiversity conservation, it entailed a set of challenges for international cooperation regarding biodiversity for food and agriculture in particular. Agrobiodiversity, the term I use to refer to the latter throughout this book, cannot be equated with biodiversity in the broader sense, which the CBD targets, and comes with specific implications for conservation that require separate attention, as the negotiations and the final text of the CBD make unambiguously clear. In the following, I delineate the developments brought about by the CBD with a focus on its ultimate significance for the global system of agrobiodiversity conservation.

An important development the CBD consolidated was a discursive shift from "plant genetic resources" to "biodiversity" that had been underway since the Brundtland report *Our Common Future* (WCED 1987).[30] With the CBD,

30 The Brundtland Commission is the informal name for the World Commission on Environment and Development (WCED), derived from its chair, the former Norwegian Prime Minister Gro Harlem Brundtland. The UN created the WCED in 1983 as an independent body engaging with policy issues in the interplay of environment and devel-

"biological diversity" superseded the overtly utilitarian notion of genetic resources as the core reference for conservation efforts. Biological diversity and its short form biodiversity more broadly denote "the variability among living organisms from all sources including, *inter alia*, terrestrial, marine and other aquatic ecosystems and the ecological complexes of which they are part: this includes diversity within species, between species and of ecosystems" (Secretariat of the CBD [1992] 2011: 4, Art. 2). This understanding introduced a new ecological framing insofar as it replaced the compartmentalisation of species with an ecosystemic and processual understanding of biological diversity. The CBD also paved the way for discussions around "biocultural" diversity by linking biological diversity to local and Indigenous knowledges, innovations, and practices worthy of protection alongside biological materials (Secretariat of the CBD [1992] 2011: 8, Art. 8j; see also Fenzi/Bonneuil 2016: 78). All this attracted renewed attention to *in situ* conservation approaches even from previously predominantly *ex situ*-oriented actors. The IBPGR, as of 1991 active as the International Plant Genetic Resources Institute (IPGRI),[31] for instance, launched a global project on "Strengthening the Scientific Basis of In Situ Conservation of Agricultural Biodiversity On-Farm" in 1997 (see Jarvis/Hodgkin 1998).

Fenzi and Bonneuil (2016: 78) argue that the proliferation of the biodiversity paradigm with its connotations of ecological complexity, interdependence, and resilience did not, however, undo what they call a "resourcist cosmovision". While the conception of nature became discursively more holistic, advances in genetic technologies allowed for unprecedented forms of cross-species breeding, leveraging biological diversity as a reservoir of genetic resources (Fenzi/Bonneuil 2016: 78–80; see also Flitner 1995: 233).[32] Fenzi and Bonneuil

opment. It dissolved with the publication of the Brundtland Report, which famously established the principle of "sustainable development" informing UN policy on environment and development to this day (Gómez-Baggethun 2019; see also Flitner 1995: 231).

31 In 2006, the IPGRI, formerly IBPGR, was renamed again, this time as Bioversity International "in an apparent effort to stress the importance of biodiversity as the founding concept" (Fenzi/Bonneuil 2016: 78).

32 Pellegrini and Balatti (2016: 2760–2761) further discuss how the emergence of the biodiversity concept was also associated with the environmental movements emerging in the 1960s, which influenced public opinion and contributed to the proliferation of a 'new environmental paradigm' and 'ecocentric vision' of nature as something to be pro-

attribute this development to the dominant position of the biotechnology industry within the international political arena and trade regime. Although the FAO attempted to counteract the dominance of the "new CBD/WTO regime" through "a more mutualistic logic aimed at bringing together the positions of breeders, developing countries, and NGOs" (Fenzi/Bonneuil 2016: 80), it never really queried the resourcist approach to the natural world as a reservoir of *plant genetic resources for food and agriculture* – a term that gained a foothold with the developments in the 1990s (see FAO 1996a, 1997, 2001). The agricultural sector did not follow the CBD's discursive shift from plant genetic resources to biodiversity for many years to come, as is discernible from the denomination of the FAO's periodically compiled report on *The State of the World's Plant Genetic Resources for Food and Agriculture* (FAO 1997, 2010). It was only recently renamed as *The State of the World's Biodiversity for Food and Agriculture* (FAO 2019). The anthropologist Thom van Dooren (2009b: 105) argues that this conceptual divergence of the CBD with its focus on biodiversity and the agricultural sector's discourse on PGRFA indicates the clear interest of the latter in genetic material as bearer of certain genetic traits and their conservation for use.

Another discursive shift that the CBD effected, in contrast, did strongly affect the agricultural sector and made negotiations of further agreements necessary. The professed and primary objective of the CBD was to ensure "the *conservation* of biological diversity, the *sustainable use* of its components and the *fair and equitable sharing of the benefits* arising out of the utilization of genetic resources, including by appropriate *access to genetic resources* and by appropriate *transfer of relevant technologies*" (Secretariat of the CBD [1992] 2011: 4, Art. 1, emph. added). The Convention not only assigns the *responsibility* for fulfilling these objectives to nation-states, but also legally enshrines the *sovereign right* of nation-states to exploit the resources occurring within their territory in Article 3 of the CBD. Observers highlight that, by doing so, the CBD (following the 1991 FAO resolution amending the IU) paved the way for the "national appropriation of biological resources (particularly of genetic resources), shifting the notion of *heritage of mankind* to *heritage of nations*" (Pellegrini/Balatti 2016: 2757, emph. in orig.; see also Visser 2012: 267).

By thus meeting a long-standing demand voiced by countries in the Global South seeking to protect resources originating in their national territory against breeders' unrestricted access and intellectual property claims, the

---

tected from man, replacing the 'dominant social paradigm' and anthropocentric vision of nature as a resource by the end of the 1970s.

CBD contributed to an atmosphere of conciliatory cooperation among the contracting parties and other non-governmental actors involved (see Fenzi/ Bonneuil 2016: 78). At the same time, it provisionally diminished the prospects of establishing a global safety duplication seed bank such as envisioned in the SIS proposal because the predominant matters of dispute were now questions of access and benefit sharing rather than long-term conservation (Qvenild 2008: 111). What is more, the CBD's provisions created administrative barriers for all regarding international exchange of PGRFA. To unfold these changes and the ensuing developments, I shall now take a step back to delineate the development of the CBD.

The first initiative to develop a convention on biodiversity within the framework of the UNEP came from the US State Department in 1987, at the peak of the FAO's controversy about rights over PGR. According to Flitner (1995: 239–240), a first draft proposed for discussion by the International Union for Conservation of Nature and Natural Resources (IUCN) in 1988 shows overlaps with the debates in the FAO, especially with respect to the facilitation of free access to PGR in exchange for financial compensation. He notes that the USA may have hoped to establish a framework that would be more in their interest than what seemed to be underway in the FAO. However, through the Group of 77, the 'developing countries' made a strong case for the preservation of the principle of national sovereignty over PGR, which was ultimately enshrined in the CBD, albeit in a narrower formulation under pressure from 'developed countries', first and foremost the USA.

The final text of the convention ensures "[national] authority to determine access to genetic resources" (Secretariat of the CBD [1992] 2011: 11, Art. 15.1) only with respect to "those [genetic resources] that are provided by Contracting Parties that are countries of origin of such resources or by the Parties that have acquired the genetic resources in accordance with this Convention" (Secretariat of the CBD [1992] 2011: 12, Art. 15.3). Consequently, plant genetic material already collected and banked in the IARCs' seed repositories, for instance, fell outside the jurisdiction of the CBD (Flitner 1995: 239–242). In addition, the stipulation that national sovereignty over collections of plant genetic resources is acknowledged if and when a country is the country of origin of the PGR in the respective collection (or has acquired them in accordance with the terms of the CBD) posed a particular difficulty for the agricultural sector:

> "For wild resources it is normally possible to identify more or less their point of origin, and also when there is more than one country of origin. However,

> domesticated resources have been developed gradually via exchanges between farmers and breeders over short and long distances for hundreds and thousands of years, in several cases since the beginnings of agriculture. Determining the countries of origin of these resources [...] is in most cases virtually impossible." (Andersen 2003: 45–46)

In today's transnationally and transregionally interdependent agricultural system, a bilateral principle of granting access and sharing benefits, as enforced by the CBD, thus did not facilitate but rather complicated the global exchange of PGRFA. A recognition of the need for further negotiations factoring in the particularities of PGRFA was therefore included in the CBD and delegated to the FAO and the CPGR (see also Manzella 2012: 151).

According to Esquinas-Alcázar et al. (2012: 141), the CBD's lack of awareness with regard to the specific needs of the agricultural sector arises from the underrepresentation of the latter in the negotiations of the convention. It was only at the Conference for the Adoption of the Agreed Text of the CBD held in Nairobi in May 1992, that a group of representatives of the agricultural sector issued a draft resolution on agricultural biodiversity. It was incorporated into the Nairobi Final Act as Resolution 3, "The Interrelationship Between the Convention on Biological Diversity and the Promotion of Sustainable Agriculture" (Secretariat of the CBD 2005: 406–408), and adopted by all signatories to the CBD along with it. This resolution recognised the need to bring the enduring debates on access to *ex situ* collections acquired in a manner nonconforming to the CBD to a favourable conclusion for all parties involved along with the debates on farmers' rights (Secretariat of the CBD 2005: 408). Furthermore, it recommended strengthening the Global System for the Conservation and Sustainable Utilization of Plant Genetic Resources for Food and Agriculture that the FAO, IPGRI, and the CGIAR pursued (which consisted only of the IU at that time) through a global plan of action that should be based on an assessment of the state of the world's plant genetic resources (Secretariat of the CBD 2005: 407).

In the following years, the FAO conducted a global assessment project in more than 150 of its member countries and appointed Cary Fowler to lead it. It resulted in the first *State of the World's Plant Genetic Resources for Food and Agriculture* (FAO 1997). This report was unprecedented, as the political scientist Regine Andersen (2003: 49, 2008: 98–99) notes, both in its extensiveness, insofar as it covered not only biological but also technical and institutional aspects, and in the way it prompted conservationists and politicians to take action. In the same period of time, based on these assessments and also under the direction of Fowler, the FAO negotiated a Global Plan of Action for the Conservation and Sustainable Utilization of Plant Genetic Resources for Food and Agriculture (FAO 1996a; hereafter: GPA for Global Plan of Action) that was adopted in 1996 at the fourth International Technical Conference on Plant Genetic Resources in Leipzig, Germany.[33] The Leipzig Declaration (FAO 1996b), which announced the results of the conference and adopted the GPA, emphasised the importance of conserving and sustainably using PGRFA as well as of sharing the benefits arising from their use. Both the GPA and the FAO's report on the state of the world's PGRFA were designed to be periodically updated in order to serve as a means for the CPGR to continuously "recommend priorities and to promote the rationalization and coordination of efforts" (FAO 1996a: 12). Their purpose was to help the FAO member states "monitor activities for the conservation and utilization of plant genetic resources, determine priorities, and oversee programmes in this area through the Commission" (FAO 1996b: 3).[34]

As for the pending regulatory issues around PGRFA raised in Resolution 3 of the Nairobi Final Act of the CBD, the FAO and the Conference of the Parties to the CBD (COP) agreed that they should be resolved in harmony with both the CBD and the IU. The FAO mandated the CPGR to oversee the revision of the IU (Visser 2012: 267). The endeavour was tricky insofar as it meant harmonising two fundamentally divergent principles: "The International Undertaking was born out of a need for *facilitating access* to genetic resources, whereas the CBD emphasised the need for *fair and equitable sharing of benefits* arising out of the use of genetic resources" (Andersen 2003: 45, emph. added). However, new international trade agreements developed in the following years increased the urgency of the need for a legally binding successor agreement to the merely voluntary IU – first and foremost the World Trade Organization's (WTO) Agreement on Trade-Related Aspects of Intellectual Property Rights (TRIPS Agreement) passed in 1994, which committed the contracting parties to ensure some form of protection for plant varieties and effectively strengthened

33 After the first Technical Conference on PGR, which was co-organised with the IBP in 1967, the FAO had convened a second and third Technical Conference on PGR in 1973 and 1981.

34 The FAO published a second edition of the State of the World's PGRFA in 2010 and a second GPA in response to it in 2011. In 2019, the third edition of the State of the World's PGRFA appeared, albeit with a terminological shift from PGR to biodiversity (see FAO 2019).

the UPOV absent of another "effective *sui generis* system" (Andersen 2003: 44). Thus arose "what seemed unimaginable just a little while ago – namely a united front of developed and developing countries, the seed industries and NGOs, working towards a common political objective" (Esquinas-Alcázar/Hilmi/López Noriega 2012: 142). Part of the envisioned agreement, next to strengthening the agricultural sector's position vis-à-vis the commercial and the environmental sector, was to ensure the long-term conservation of agrobiodiversity and make it fairly and equitably accessible for research and breeding, as spelled out in the GPA. In 1994, the by then twelve IARCs of the CGIAR that operated a genebank, altogether holding almost 450,000 samples, placed their collections under the patronage of the FAO, thus making them available to all FAO member countries for agricultural use, breeding, and research (Andersen 2003: 48).

In the very same year, formal negotiations began on what would become, over the following seven years, the International Treaty on Plant Genetic Resources for Food and Agriculture (ITPGRFA; hereafter: Seed Treaty; also referred to as the Plant Treaty or simply the Treaty within the agrobiodiversity conservation community).[35] According to Esquinas-Alcázar, who was a key figure in the negotiations on the Seed Treaty in the role of secretary of the CPGR (renamed as the Commission on Genetic Resources for Food and Agriculture (CGRFA) as of 1995), the debates of the newly united agricultural sector took place in a "highly constructive atmosphere" (Esquinas-Alcázar/Hilmi/López Noriega 2012: 142). Esquinas-Alcázar et al. further describe the climate in which the Treaty was ultimately adopted as one "of widespread euphoria" (2012: 143). External observers paint a much more reserved picture. Andersen, a long-term observer of the Seed Treaty and its emergence, notes that the negotiations were "extremely difficult, core provisions in the text were in brackets right up to the last bell, and it seemed impossible to unite all fronts on joint solutions" (2003: 50). According to a report by the ETC Group, Esquinas-Alcázar and Clive Stannard, who represented the CGRFA Secretariat in the negotiations, were often "the only two people in FAO who believed in the treaty" (ETC Group 2001: 3). After the Treaty was adopted, the report says, appreciation of the two "went well beyond courtesy to something akin to cult worship" (ibid.).

---

35 Fowler, who had previously been working for the FAO in various capacities, represented the CGIAR in the negotiations on the Treaty.

The negotiations on the Seed Treaty lasted throughout three regular and six intermediate meetings of the CGRFA as well as several meetings of a specially founded contact group between 1994 and 2001 (Andersen 2003: 50; Visser 2012: 266). Unable to reach consensus, in November 2001 the FAO put the agreed text of the Treaty to a vote. With the USA and Japan abstaining from the vote, an overwhelming majority of 116 countries adopted the Treaty. 90 days after the first 40 countries had ratified, accepted, or approved it, as determined within the Treaty, it entered into force in June 2004 and became operative with the first session of its Governing Body in Madrid, Spain, in June 2006.[36] The Seed Treaty provides the "first legally binding agreement exclusively pertaining to plant genetic resources for food and agriculture" (Andersen 2003: 50) and implements the goals formulated in the GPA of 1996. Article 1 of the Treaty, which states its objectives, commits all signatories to ensuring "the conservation and sustainable use of plant genetic resources for food and agriculture and the fair and equitable sharing of the benefits arising out of their use" (FAO 2001: 2). What distinguishes the Seed Treaty from the CBD, at its core, is that the latter is enforced in and through bilateral arrangements, whereas the Treaty established a multilateral system (MLS) of access and benefit sharing. It facilitates easier access to and exchange of PGRFA only for the specific purpose of "utilization and conservation for research, breeding and training for food and agriculture" (FAO 2001: Art. 12.3(a)). Other purposes such as "chemical, pharmaceutical and/or other non-food/feed industrial uses" (Manzella 2012: 154) are not covered by the regulations of the MLS and require further bilateral negotiations between the contracting parties.

The MLS legally acts through a Standard Material Transfer Agreement (SMTA) between the requester and the provider of plant genetic material shared under the MLS. Through the SMTA, any transaction including PGRFA processed by an institution under the jurisdiction of a country that has signed the Treaty accepts and realises the conditions of the MLS (Manzella 2012: 151–152). Importantly, this only applies to a limited range of crop plants the parties involved in the negotiations of the Treaty agreed to share through the MLS, and these are listed in Annex 1 to the Treaty. Furthermore, the MLS only affects those PGRFA "'under the management and control' of the Contracting

36 When it first came into operation, a total number of 105 countries as well as the European Union had ratified the Treaty (see Andersen 2003: 50; Esquinas-Alcázar/Hilmi/López Noriega 2012: 143–144). As of 1 April 2025, the membership list of the Treaty comprises 154 countries and the EU as an intergovernmental signatory (see FAO 2025).

Party and those that are 'in the public domain'" (ITPGRFA Art. 11.2 as quoted by Manzella 2012: 153), that is, those untouched by intellectual property rights.

Annex 1 specifies 35 food crops and 29 forage crops (FAO 2001: 46–49). It "includes important food crops such as rice, wheat, maize, rye, potatoes, beans, cassava, and bananas. But it excludes species such as soybean, oil palm, cotton, sugar-cane, cocoa, groundnut, most vegetables, and important tropical forage plants." (Andersen 2003: 51) While it is usually emphasised that the 64 crops in Annex 1 make up a high proportion of the world's most important food crops for human nutrition, the composition of this list is the result of a long and difficult negotiation process. Bert Visser highlights the political and economic interests that shaped these negotiations and their outcome:

> "Regions objected to the inclusion of certain crops not because these crops were considered to be unimportant for food security but, rather, because the genetic resources of these crops were considered to be, first and foremost, economic assets for those particular regions. This is how soybean and groundnut, sugar cane and oil palm disappeared from the list, even though they had been featured for a considerable period. It also explains why some crops that no country would seriously consider to be of utmost importance to food security are included on the list, such as asparagus and strawberries. Ultimately, no region objected to a proposal to include these crops, and so they stayed on the list." (Visser 2012: 279)

International exchange of PGRFA not included in the Treaty is regulated by the CBD if these resources were acquired after the CBD entered into force (of course, this only applies to its contracting parties). No PGRFA acquired earlier fall under any binding international regulatory framework (Andersen 2003: 51).

Next to the conditions of access, the SMTA also regulates the sharing of benefits gained from the utilisation of material received through a transaction under the MLS, for example from the commercialisation of new varieties bred with the material in question (Esquinas-Alcázar/Hilmi/López Noriega 2012: 144–145). Whenever PGRFA listed in Annex 1 are shared by contracting parties of the Seed Treaty, they are officially considered as being provided by the pool of plant genetic material shared under the MLS. Accordingly, all benefits derived from these resources are to flow back into the MLS and be equally shared. While the Treaty also identifies non-monetary modes of benefit sharing, such as "the exchange of information, the access to, and transfer of, technology, and capacity building" (Manzella 2012: 155), the central goal is the sharing of monetary benefits through a multilateral Benefit Sharing Fund (BSF). The Treaty stipulates that the resources in this fund are to be directed primarily towards the support of farmers in the Global South for the conservation and sustainable use of PGRFA (Manzella 2012: 155–157). At the time of the adoption of the Treaty, the observant NGO GRAIN (2001) critically remarked that the amount, form, and conditions of payments to the BSF were still unclear.[37] The first obligatory payment was made in 2018 by a Dutch breeding company that had previously acquired and profitably processed resources from the MLS through an SMTA. According to the director of science of the Crop Trust, however, this delay is not necessarily politically motivated insofar as the breeding and marketing of new varieties can take a full decade and therefore was not to be expected much sooner (see below). As of 2019, "the Fund has supported over 60 projects in 55 developing countries, using USD 20 million in voluntary contributions" (Crop Trust 2019; see also Guarino 2018).

While the Seed Treaty and the MLS have successfully linked the facilitation of global flows of PGRFA to the acknowledgment and protection of the contribution made by farmers as the main providers of PGRFA, which is explicitly recognised in Article 9 of the Treaty, farmers' rights remain one of the core issues of debate today. This goes back to the rather unspecific recognition in the Treaty of their contribution to producing and conserving PGRFA and the need to protect their rights. NGOs criticised this from the beginning, saying that it "leaves the responsibility for implementing these rights to national governments and is not backed up by any international mechanism or enforcement procedure" (GRAIN 2001; see also ETC Group 2001: 6). The failure to implement the Treaty's provisions on farmers' rights eventually led to the creation of an intergovernmental Ad Hoc Technical Expert Group (AHTEG) for the assessment and realisation of farmers' rights in 2017. This group meets annually with the aim of monitoring and making informed recommendations for improving the implementation of Farmer's Rights as set out in Article 9 of the Treaty.[38]

37 GRAIN is an independent international non-profit organisation that was founded in 1990. It emerged out of a group of international activists engaged in research, advocacy, and lobby work to raise awareness about genetic erosion in agricultural biodiversity in the 1980s. Today, GRAIN "support[s] small farmers and social movements in their struggles for community-controlled and biodiversity-based food systems" (see GRAIN n.d.).

38 See http://www.fao.org/plant-treaty/areas-of-work/farmers-rights/expert-group/en/ (last accessed July 18, 2025). In 2018 the UN's Human Rights Council adopted the United Nations Declaration on the Rights of Peasants and Other People Working in

Another core controversial issue in the negotiations on the Treaty, which was the main reason for the USA and Japan to abstain in the final vote, is the question of intellectual property rights (see ETC Group 2001: 4). The majority of the negotiating parties agreed that recipients of genetic material shared under the MLS may not "claim any intellectual property or other rights that limit the facilitated access to the plant genetic resources for food and agriculture, or their genetic parts or components, in the form received from the Multilateral System" (FAO 2001: Art. 12.3(d)). GRAIN's report on the adopted version of the Treaty points out the ambiguity of this formulation: "While optimists will say it leaves room for an anti-IPR stance, the article in question basically states that the seeds and other genetic materials governed by the Treaty can be patented – as long as they are modified in some way." (GRAIN 2001)[39] Notwithstanding the shortcomings and critical reservations mentioned here, NGOs and CSOs for the most part agreed that

> "the legally binding treaty, with an access list including most of the world's major food crops, offers a good platform upon which to rebuild international cooperation and public sector agricultural science. In the years ahead, the treaty will be a platform to be proud of as long as governments ratify it quickly and as long as the FAO realizes it has a horse it should run with." (ETC Group 2001: 3)

With the Seed Treaty, an inter- and more-than-governmental platform was created for debate on and the development of policies related to the conservation and sustainable use of PGRFA.

What is more, as the first legally binding agreement settling many of the heatedly debated issues in international agrobiodiversity politics, at least to a certain degree, most importantly access to and exchange of PGRFA, the Seed Treaty was a key step on the way towards successfully establishing an international backup seed bank:

> "The idea of having a secure global backup system for hundreds of individual gene banks [...] is fundable only if the gene banks agree to cooperate

---

Rural Areas thus strengthening the case for realising farmers' rights as stipulated by the Treaty (see UN Human Rights Council 2018).

39 Since the adoption of the Treaty, the issue has continued to re-emerge in various forms, the latest of which is an ongoing heated debate on digital sequence information patenting (see Nehring 2022).

> and to address more than their immediate national concerns: that is, they must agree to a common standard of access and benefit sharing. The international treaty provided this standard, making it possible to think of constructing a communal safety net for all crop diversity held in gene banks." (Fowler 2008b: 190)

Although no particulars were stated at the time, the Treaty established the promise to develop a funding strategy for the implementation of an international PGRFA repository (Andersen 2003: 51). In 2002, during the World Summit on Sustainable Development – the Second Earth Summit – in Johannesburg, South Africa, the FAO and the CGIAR through IPGRI formally committed to establishing the Global Crop Diversity Trust (today: Crop Trust). Its core responsibility would be to manage a Crop Diversity Endowment Fund to permanently and self-sustainingly support the global system of conservation (see Crop Trust 2025). Formally established by the FAO in 2004, with Fowler appointed inaugural director in 2005 (a position he held until 2012), the Crop Trust convened its first meeting in 2006. At this meeting, a relationship agreement with the Governing Body of the Seed Treaty was signed recognising the Crop Trust "as an essential element of the Funding Strategy of the International Treaty in relation to the *ex situ* conservation and availability of plant genetic resources for food and agriculture" (FAO 2006: Art. 2). With this financial scheme and the regulatory framework of the Treaty dispelling any remaining doubts about property rights, the way was finally clear for a second, eventually successful initiative to establish an international seed bank in Svalbard, which would act as a global safety net for seed collections all over the world: the Svalbard Global Seed Vault.

1.5 The Svalbard Global Seed Vault (I): Safeguarding the World's Plant Genetic Diversity

With both the Seed Treaty and the Global Crop Diversity Trust up and running, IPGRI approached the Norwegian government a second time with the suggestion to establish an international seed storage facility in Svalbard for safely storing duplicates of genebank collections worldwide. The Norwegian Ministry of Foreign Affairs, together with the Ministry of Agriculture, commissioned a committee of experts to conduct a Study to Assess the Feasibility of Establishing a Svalbard Arctic Seed Depository for the International Community (Fowler et al. 2004).[40] With the declared goal of establishing "an ultimate safety net" as a "security against 'worst case scenarios'" (Fowler et al. 2004: 2), the study aimed to assess "whether a facility located in Svalbard might provide ultimate 'failsafe' protection for the world's most valuable natural resources, and whether it might be able to do so in a manner that is efficient, sustainable, inexpensive, and politically and legally acceptable" (Fowler et al. 2004: 1). The following requirements were carefully evaluated: whether such a facility could (a) be designed to last forever at best, but at least for 200 years; (b) meet the highest international long-term conservation standards; (c) provide sound protection against all kinds of natural and human-induced disasters; (d) receive international political and financial support; (e) be operated at fairly low costs; and whether it would (f) be used as a backup repository by a majority of the existing holders of seed collections (Fowler et al. 2004: 2).

The feasibility study determined that Svalbard provides the safest possible environment to be found in the world for establishing a secure long-term conservation repository, on the grounds of four main factors. First, due to the remoteness of Svalbard and the fact that it belongs to Norway, which has a reputation for being politically neutral and stable, the risk of threats arising from political instability, war, or terrorist attacks is considered very low. Second, the archipelago rates as geologically safe due to a lack of tectonic activity in the Arctic, which reduces the risk of earthquakes and associated threats.[41] Third, the Arctic's cold temperatures and permafrost are considered to offer unique conditions for long-term conservation purposes. Insofar as sub-zero temperatures decelerate biochemical processes of life and decay, they facilitate the preservation of biological materials (see e.g. Friedrich/Höhne 2016). Although the FAO's *Genebank Standards for Plant Genetic Resources for Food and Agriculture* (2014: 24)[42] require a storage temperature of -18 degrees Celsius to guarantee

---

40 The committee was chaired by Fowler, at that time Director of Research of the Center for International Environment and Development Studies at the Agricultural University of Norway and Senior Advisor to the Director General of the IPGRI. For an overview of the committee members see Fowler et al. (2004: viii [Annex V]).

41 A critical commentary published by GRAIN shortly after the opening of the Seed Vault challenges this aspect: "Just days before the opening of the Vault, Svalbard was at the centre of the biggest earthquake in Norway's history, even though the facility's feasibility study assured that '*there is no volcanic or significant seismic activity*' in the area" (GRAIN 2008: 2, quoting Fowler et al. 2004: 4, emph. in orig.). For further information on the seismic activity mentioned here see Pirli et al. (2010).

42 The first version of these *Genebank Standards* was published by the IBPGR in 1985.

long-term viability, the Arctic permafrost's year-round natural average temperature of at least -3 degrees Celsius would guarantee continued cold storage for a storage facility built into the permafrost even in case of a temporary failure of the technical refrigeration system (see Fowler 2008c: 19). The availability of local coal mines able to provide a source of power for the technical cooling system would keep the cost of the facility's maintenance low and mean easy maintenance. Fourth and finally, building a vault in Svalbard's mountainscape with an entrance high above sea level would secure the seeds against floods and rising sea levels, thus making it a suitable location even in the face of dangers resulting from global warming.

Based on the positive results of the feasibility study and the international agrobiodiversity conservation community's endorsement of the proposal, by the end of 2004 the Norwegian government had committed to building an international seed storage facility near the town of Longyearbyen on the Arctic island of Spitsbergen in the Svalbard archipelago. On February 26, 2008, the Svalbard Global Seed Vault was officially opened, which was celebrated with a ceremony for storing the first more than 320,000 seed samples (Norwegian Ministry of Agriculture and Food n.d. a). The Seed Vault is cooperatively run by a tripartite partnership between the Norwegian government, the Crop Trust based in Bonn, Germany, and the Nordic Genetic Resource Centre (NordGen) based in Alnarp, Sweden. As legal owner of the facility, the Norwegian government bears the overall responsibility. Funding for daily operations and management as well as staff for surveillance and monitoring are provided by the Norwegian Ministry of Agriculture and Food through Statsbygg, the Norwegian public sector administration enterprise responsible for government estates. Beyond that, funding for operational costs and assistance to depositors is provided by the Crop Trust, which administers international donations. NordGen is responsible for daily management operations such as consulting with seed banks about conservation options, preparing documents, and facilitating shipments, as well as receiving, storing, and documenting seed depositories. An International Advisory Panel (IAP) composed of seven people representing genebanks, plant breeders, and other stakeholders, including a secretary delegated by NordGen oversees the operations and management of the Seed Vault.

What essentially distinguishes the Seed Vault from other conservation facilities for PGRFA is that it is not a genebank in the narrower sense. Whereas genebanks have active and working collections used by farmers, breeders, and scientists for immediate or medium-term purposes, the purpose of the Seed Vault is to provide an internationally accessible secure storage space for the long-term conservation of dormant duplicates of these kinds of collections (FAO 2014). Pellegrini and Balatti (2016: 2761–2765) distinguish three seed bank profiles: "Assistentialist", often community-based seed banks store and provide seeds directly to local farmers, usually with an interest not in long-term conservation but in facilitating exchange and independence of private seed companies. "Productivist" seed banks are focused on research and breeding to further improve the production and economic value of locally relevant varieties. "Preservationist" seed banks such as the Svalbard Global Seed Vault serve as backup repositories for duplicates of other collections with the primary purpose of long-term conservation and availability of the highest possible crop genetic diversity. While genebanks, especially those with a productivist profile, are usually complex infrastructures with sophisticated technical equipment and trained staff, the Seed Vault is designed as a low-tech, low-maintenance, and low-cost facility that requires no permanent staff.

It is frequently, but mistakenly, assumed that the Seed Vault was built in a former coalmine on Svalbard. While the backup of the Nordic seed collection, which had inspired the idea of a Global Seed Vault in Svalbard, was previously stored in an abandoned coal mine, the feasibility study for the Seed Vault determined that the high level of hydrocarbon gases in these mines would be too much of a safety hazard. Instead, the Seed Vault was carved into the "virgin rocks" (Norwegian Ministry of Agriculture and Food n.d. a) of Svalbard's Platåberget (Norwegian for "plateau mountain"), which contains no coal deposits. The entrance to the facility lies halfway up the mountain at 130 metres above sea level. It is marked by a concrete block adorned with an artwork, like every government building in Norway since the Art Programme for New Government Buildings was launched in 1997.[43] The mosaic-like installation on the top of the entrance element emanates a turquoise light, which has made for spectacular pictures especially in the dark of the long Arctic winter. It was designed by Norwegian artist Dyveke Sanne and is called "Perpetual Repercussion". According to Fowler,

> "[through] hundreds of small triangular pieces of polished steel with a mirrorlike quality [...] the light sculpture reflects the environment, as the seeds reflect theirs. But in Dyveke Sanne's view it is more than that. 'The thought or idea of the installation is precisely to insist on reflection, that who you will

---

43 For more information on this programme, see https://koro.no/english/a-short-story-about-koro/ (last accessed July 18, 2025).

> meet in the mirror is yourself and that whatever needs doing is up to you.'" (Fowler 2016: 121)

Behind the entrance door, a tunnel leads 120 metres deep into the mountain where 40 to 60 metres of solid rock and permafrost cover three seed storage chambers. Each of the three storage halls inside the vault has the capacity to store 1.5 million seed samples. A sample, also called an accession, contains an average of 500 seeds, meaning that the Seed Vault can store up to a total of 2.25 billion individual seeds. As of July 3, 2025, the Seed Vault contains 1,356,591 seed samples representing 6,378 species stored by 129 depositors (NordGen n.d. b) thus exhausting less than a third of its capacity. There are three main prerequisites to making a deposit in the Svalbard Global Seed Vault. First, the mandate of the Seed Vault only covers PGRFA. Second, all seed samples stored in the Vault must be original, meaning that no identical accession can be stored by the same or another genebank. Third, depositors agree to facilitate access to those resources in their genebanks which are duplicated in Svalbard (Westengen et al. 2020: 1315). Depositors recognise these terms by signing an agreement with the Norwegian Ministry of Agriculture and Food, which stipulates all rights, responsibilities, and obligations of both parties with respect to the deposited material and the conditions of storage and withdrawal (see NordGen 2008). This agreement also contains the provision on property rights remaining with the depositor that was crucial before many genebanks, especially in the Global South, would agree to store duplicates of their collections in the Seed Vault in the first place. Storing duplicates in the Seed Vault is free of charge, but the depositor agreement commits depositors to cover the costs of packaging and shipping. If required, the Crop Trust provides financial and technical assistance.

With no permanent staff based in Svalbard, seeds are usually brought into the Seed Vault three times a year on pre-scheduled occasions. The Seed Vault coordinator and technicians who handle the deposits who are employed by NordGen travel to Svalbard for these occasions. Seed shipments usually reach Longyearbyen by airfreight. Before entering the Seed Vault, they are x-rayed for security purposes at the local airport to make sure that nothing enters the facility except seeds. A "'black box' arrangement" (Fowler 2008b: 190; see also Asdal 2025) prohibits anyone except the depositor from opening the boxes, since they remain property of the depositor at all times. All boxes that go into the Seed Vault must fulfil certain standards with respect to size and material in order to fit the shelves inside the storage chambers. They are tagged with a label providing information on the depositor and content. The seeds inside the boxes have to be dried and vacuum-packed in specifically designed aluminium pouches to meet the conditions for long-term cold storage as specified in the FAO's *Genebank Standards* (2014). Depositors further commit to providing an electronic inventory of their deposits, which contains detailed information on the depositing institution, the seed collection and its origin, the year of harvest (which identifies the regeneration cycle of a seed sample), as well as information on the specific accessions including, for instance, the number of seeds in a sample. All this information is openly accessible through the Seed Portal, an online database managed by NordGen (see NordGen n.d. b). Pictures of the Seed Vault, its construction, storage events, and more are accessible through the Svalbard Global Seed Vault's and the Crop Trust's accounts on flickr. The Seed Vault's website also offers a virtual tour of the facility (Norwegian Ministry of Foreign Affairs/Crop Trust/NordGen n.d.).

Throughout the chapters that follow, I analyse the ways in which the Svalbard Global Seed Vault is a nodal point of the historically emergent and spatially distributed world of agrobiodiversity conservation. Rather than seeing the Seed Vault as an institutional addition to the international agro-political arena, I primarily consider it as itself a political arena within the world of agrobiodiversity conservation. This enables me to explore the agrobiodiversity politics enacted by, through, and around the Seed Vault. The following chapter shows how this analytical perspective expands the state of social-scientific and humanities research on the Seed Vault and *ex situ* agrobiodiversity conservation (2.1), and delineates the empirical-analytical process underlying this study including its methodological premises (2.2).

Figure 1: A Short History of International Agrobiodiversity Conservation Efforts (the author, 2025)

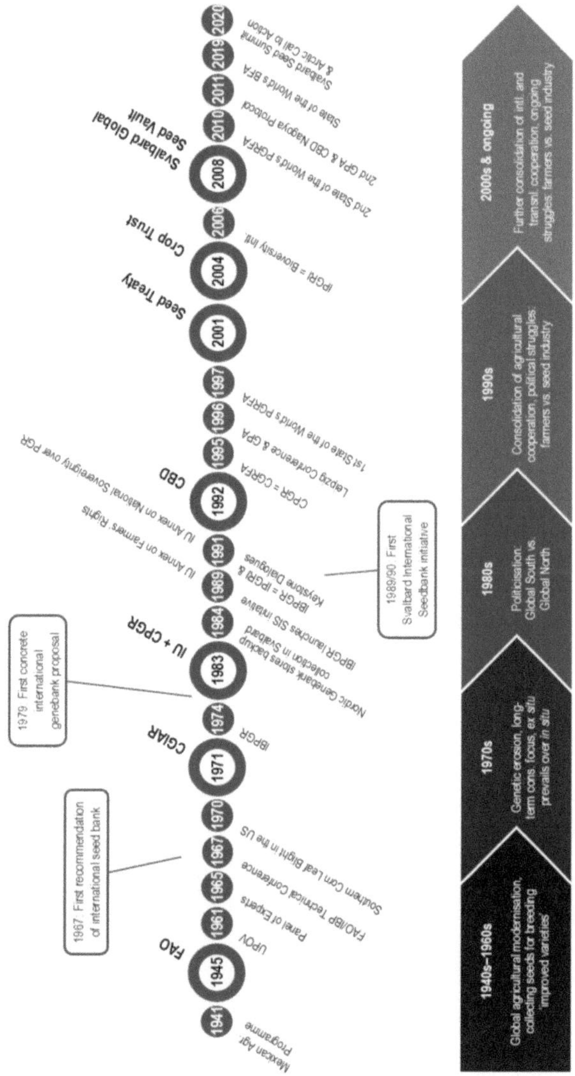

2. The Svalbard Global Seed Vault and Agrobiodiversity Conservation in Social-Scientific Research

The empirical research underlying this study, as the book's subtitle states, is an inquiry into the world of agrobiodiversity conservation *through* the Svalbard Global Seed Vault. This means that the Seed Vault is not the object so much as the starting point and nodal point – or the seed – of my research. As unfolded in the previous chapter, it is the product of historical and ongoing political, economic, technological, and material-discursive struggles and processes. Throughout the chapters that follow, I show that since its establishment in 2008 it has become a new agro-political actor and arena assembling other actors participating in shaping how agrobiodiversity loss and conservation is known, imagined, performed, and transformed. Put differently: like the seeds it assembles, the Seed Vault comes from and comes with a world that has created it while also becoming what it is with the world(s) it assembles and intervenes in. It is on these world(s) and world-making processes, in which the Seed Vault is entangled, that I focus my attention in the ensuing discussions of what the Seed Vault is and does. In a first step, this chapter develops this analytical perspective as a desideratum in social-scientific and humanities research on the Seed Vault and *ex situ* agrobiodiversity conservation. In a second step, it elaborates the empirical-analytical research process of my study, which is guided by the methodology of Situational Analysis.

2.1 The Current State of an Emergent Field of Research

In the period of less than two decades since the Svalbard Global Seed Vault first opened, it has attracted a notable amount of public attention in the form of countless international press reports, documentary films (e.g. *Seed Warriors* [2009]; *Seeds of Time* [2013]; *Seed: The Untold Story* [2016]) as well as in fiction (e.g. Maja Lunde's *The Dream of a Tree*[1]) and on TV (e.g. *Seed: Deadly Power*[2]). Academic interest in the Seed Vault has so far remained scarce outside of plant genetics, conservation biology, and related fields of scientific research actively involved in *ex situ* agrobiodiversity conservation efforts. One can relatively quickly gain an overview of the existing English-language accounts of the Seed Vault in the social sciences and humanities.

This limited interest is somewhat surprising considering that different, currently thriving areas of research actually intersect in the Seed Vault and a comprehensive analysis of its operations could make a meaningful contribution to all of these fields. In the large and growing body of research on *biobanking and cryo-conservation*, attention to the banking and cold storage of seeds is rather marginal. Similarly, in the multi-disciplinary research environment critically engaging with *conservation efforts* in a more general sense, which often focus on the conservation of animal species and ecosystems, much more attention has been paid to *in situ* than to *ex situ* conservation strategies. Finally, although there has been a surge of academic interest in *agrobiodiversity conservation in genebanks* in recent years, few of the research projects in this field focus on the Seed Vault. Among those that do, only one is grounded in an in-depth empirical study comparable to the one I have conducted (see Harrison 2017; Harrison et al. 2020: 51–152). This kind of empirical research is essential insofar as it considerably differentiates the insights one can gain based on publications, online material, and media accounts.

Against this background, my study expands the existing research on the Svalbard Global Seed Vault and contributes to the various intersecting bodies of research by bringing into focus aspects, questions, connections, and perspectives that have not hitherto been addressed or have been considered marginal. Let me start by providing a brief overview of the current state of research on the Svalbard Global Seed Vault and on *ex situ* agrobiodiversity conservation efforts more generally.

1 Originally published in Norwegian as *Drømmen om et tre* in 2022.

2 A German TV series first broadcast in 2023 and originally titled *Die Saat: Tödliche Macht*.

Research on the Svalbard Global Seed Vault

The first account of the Svalbard Global Seed Vault from within the broad field of cultural studies appeared no more than a year after the Vault had opened. The environmental humanities and extinction studies scholar Thom van Dooren interrogated the Seed Vault's (self-)conception as an insurance policy for the world's endangered agrobiodiversity, observing that "[i]n the current 'climate of loss', in which biological diversity of all kinds is disappearing […] how we do conservation has taken on a new level of importance" (2009b: 103). Van Dooren criticises the *ex situ* approach underlying the Seed Vault (and the International Seed Treaty's provisions for international agrobiodiversity conservation efforts more generally) for reducing biodiversity to genetic material or (plant) genetic resources. He argues that this not only deprioritises the ecological, biocultural, and biosocial life of seeds and plants, but also concentrates the means of adaptation to changing environmental conditions in the hands of technoscientific actors rather than farmers and agricultural communities (see also van Dooren 2009a). Without denying that banking "proxies" (van Dooren 2009b: 105; see also Breithoff 2020) of lifeforms on the brink of extinction can be valuable for mitigating ecological risks and adapting to environmental change, van Dooren argues that "no matter how dire the situation, the need to conserve diversity cannot be allowed to cover over more critical discussions about how and for whom this 'conservation' work is to be done" (2009b: 109). The questions raised at this early stage by van Dooren about the Seed Vault's insurance logic, genetic reductionism, and about what it does and for whom foreshadow the key aspects of the current state of discussion about the Seed Vault in the social sciences and humanities. In the relatively small body of multi-disciplinary research on the Seed Vault two concerns can be identified. All accounts discuss *what* it is that the Seed Vault conserves by conserving seeds, on the one hand, and/or *what for*, on the other hand, meaning what it promises to achieve through the conservation of seeds.

Since the material conserved inside the vault consists of seeds, the conservation goal is evidently defined according to a molecular understanding of life and a geneticised conservation paradigm. Fern Wickson (2016) draws attention to the fact that genetically modified crop seeds are excluded from entering the Seed Vault. Exploring the reasons for and implications of this exclusion, she develops an intriguing philosophical and socio-political discussion of "the moral status and conservation value of all the mutants and monsters we are engaged in producing" (Wickson 2016: 807). Looking at the fact that the vault's collection of biomaterial isolated from its ecological entanglements is complemented by an online Seed Portal, collecting and organising digitalised bioinformation isolated from its material substrate, Suzana Alpsancar (2016) extends the focus of analysis to the role of the information paradigm and how it shapes the Seed Vault's conservation efforts. Alpsancar argues that the creation of "copies" of certain seeds and their transformation into "digital things" alters their ontological status in a way that "their materiality is only virtually effective as an insurance policy for the originals" (Alpsancar 2016: 62). The work of heritage studies scholars Esther Breithoff and Rodney Harrison (2020; see also Harrison 2015, 2017; Harrison et al. 2020) adds to this argument about the ontological politics of conservation. Breithoff and Harrison argue that the collection of agrobiodiversity in the Seed Vault and "datadiversity" (Bowker 2000) in the Seed Portal is a practice of accumulating and generating "latent forms of biocapital" (Breithoff/Harrison 2020: 102) primarily oriented towards the future realisation of economic value. The authors point out the sad irony of this speculative form of biocapital accumulation in seed banks, namely that it "requires direct speculation upon the extinction and biodiversity loss that it is created to secure the present against" (Breithoff/Harrison 2020: 119). Annukka Paajanen develops a similar argument conceiving the Seed Vault "as a site of *salvage accumulation*, where the loss of crop diversity is translated into capitalist value" (Paajanen 2024: 22–23, emph. in orig.).

The environmental anthropologist Can Dalyan investigates the promises of such an economically oriented and selective type of conservation based on a similar understanding of *ex situ* conservation as a practice of accumulating latent capital "built squarely upon the promise and premise of continued, worldwide environmental devastation" (Dalyan 2021: 182). Dalyan argues that seed banks are technoscientific tools for bio- and geopolitics. Crucially, he contends that whereas national genebanks bank on the "(re)productive capacities and latent potentialities of seeds" (Dalyan 2021: 190) as means of nation-making, international seed conservation facilities such as the Svalbard Global Seed Vault, as well as the Millennium Seed Bank Partnership (MSBP) at the United Kingdom's Royal Botanic Gardens in Kew, deploy a universalist rhetoric that elevates them to saviours of nature and humanity at large. This rhetoric, however, glosses over "imperial histories of bioprospecting that facilitated the emergence of institutions like the MSB [sic] and their vast, global collections in the first place" (Dalyan 2021: 184).[3] The sociologist of science Tahani Nadim (2018) argues that by failing to account not only for the ecological devastations of the present but also for the imperial and colonial histories "haunting" these international seed conservation facilities, "a project such as the Doomsday Vault [...] perpetuates a logic of ruination" (Nadim 2018: 245) outside the genebank that has led to the establishment of genebanks in the first place.

Although the Seed Vault was not built to be the monolithic solution for agrobiodiversity that the narrative of the 'doomsday vault' suggests it is,[4] the technoscientific *ex situ* conservation approach it embodies is rooted in a logic of "disaster preparedness" (Grigoli 2024: 118) that deprioritises the destruction of (agro)biodiversity outside the genebank. The bio-reductionist notion of biodiversity underlying *ex situ* conservation efforts, which considers (frozen) seeds as archives of (latent) "life itself" (Dalyan 2021: 189) or of a crop's biosocial evolutionary history (Peres 2016; Breithoff/Harrison 2020: 108), is the ontological foundation for the promise to adapt to changing ecological environments through selective breeding. It is this promise, and the objective of gaining "control of evolution" (Fava 2013: 146–149), that makes seed banks accumulating the largest possible diversity of plant genetic resources for future breeding *ex situ* much more promising than efforts to conserve agrobiodiversity *in situ*, where it is subject to ongoing change and erosion due to "continuing processes of natural and cultural selection" (Breithoff/Harrison 2020: 109; see also Wolff 2020: 6).

---

3 The colonial legacies of the Royal Botanic Gardens, Kew and the MSBP are the subject of careful analysis in the works of Anna-Katharina Laboissière (2019), Xan S. Chacko (2019b, 2022b), and Marleen Boschen (2022: 154–199). Kew Gardens is one of the oldest botanical gardens in the world, and its history is deeply entangled with the British Empire's colonial expansion and colonial botany (see also Crosby 1972; Brockway 2002; Schiebinger 2004; Schiebinger/Swan 2005). The seed bank at Kew Gardens, first set up in 1980, was turned into the Millennium Seed Bank (MSB) in 2000 with the aim of establishing a collection of samples of the United Kingdom's entire flora. Under the name Millennium Seed Bank Partnership (MSBP), the project's scope was soon broadened to encompass the flora of today around 100 partnering countries around the world. Boschen and Chacko interpret this process as a "postcolonial reframing" (Boschen 2022: 159) or "rebranding" (Chacko 2019b: 2) of Kew Gardens in the depoliticised spirit of global biodiversity conservation for universal human benefit, which ultimately "strengthened ties between Kew and gardens around the world [...] that had been loosening since their colonial binds were no longer controlling the flow of plants" (Chacko 2022b: 16).

4 I elaborate this argument in chapter 3.2.

Against the background of this geneticised and technoscientific approach to agrobiodiversity conservation, a last group of accounts of the Seed Vault by and large situated in sociological and cultural STS analyse it as an allegedly politically neutral means of "salvation" from the challenges climate change poses to agricultural production (Fava 2013: 118–149); as a "security technology" that secures and thus extends the present by making the threats of agrobiodiversity loss and food insecurity reversible through seed banking (Wolff 2020; see also 2024); and as a salvationist promise of "becoming-safe" through practices of "global care" for agrobiodiversity informed by imaginaries of securitisation, fortification, and mastery (Boschen 2022: 112–153). That these heroic sociotechnical imaginaries, as Elaine Gan (2015) calls them, are entrenched in gendered power relations becomes palpable in Charlotte Kroløkke's (2019) intriguing feminist cultural analysis of the documentary *Seeds of Time* (2013), in which the Seed Vault features as a womb with a birth canal through which seeds enter to be harboured and eventually leave as promissory carriers of future life.

What all accounts of the Seed Vault summarised here illustrate are the various ways in which the "specific forms of hope and security generated through practices of banking genetic diversity" (Breithoff/Harrison 2020: 119) are entangled in economic, political, and epistemic power relations. I share the conviction that it is crucial – analytically as well as politically – to attend to the ways in which the Seed Vault is caught up in these power relations and reproduces them. At the same time, I propose and pursue an analytical approach that equally attends to cracks in these power relations, that is, to the ways in which the Seed Vault and its effects in the world are not determined by them. This approach resonates with the analytical perspective the political scientist Sheryl Breen (2015) develops in her noteworthy comparative study of Native American seed saving practices and *in situ* conservation projects, on the one hand, and the Seed Vault's contribution to *ex situ* conservation efforts, on the other hand. Foregrounding how political struggles around different approaches to agrobiodiversity conservation pivot on "divergent understandings of 'seedness' and seed ownership" (Breen 2015: 39), Breen emphasises that the Seed Vault is unique, compared to other genebanks, "in its potential ability to cross the political and cultural divide over the ownership and conservation of seeds and thereby promote the vital ecological need for both *ex situ* and *in situ* seed preservation" (ibid.).

My study foregrounds this uniqueness of the Seed Vault within the larger system of genebanks it is part of, discusses its entanglement in existing power relations with a particular focus on their hitherto underexplored onto-epistemic dimensions, and explores its potential to be(come) more than what the accounts delineated above show that it currently is. I develop an understanding of the Seed Vault as a technology that is materially and discursively entangled in, but not determined by, the world it has emerged from. Its performativity and effectiveness do not arise from the design of the Seed Vault itself, but depend on the world(s) it serves and whose conservation it facilitates. This requires extending the focus from the What and What-for or How of the Seed Vault's approach to conservation – which the existing research has carefully and compellingly scrutinised – to the world(s) it is situated in and which it serves to conserve and cultivate by way of its promise to conserve futurity. This perspective is to some extent informed by a body of research engaging with the history, politics, and practices of *ex situ* agrobiodiversity conservation beyond the Svalbard Global Seed Vault, which I briefly delineate in the following.

Research on the History and Political Economy of Seed Conservation

The most comprehensive and most widely read historical account of the political economy of *ex situ* agrobiodiversity conservation is Jack R. Kloppenburg's seminal study *First the Seed: The Political Economy of Plant Biotechnology, 1492–2000* ([1988] 2004). Kloppenburg traces the history of seed banking as an entangled history of capitalism, science, and agriculture from the commodification of seeds, through the emergence of genetics and plant biotechnology and the ways in which they facilitated the modernisation and globalisation of agriculture, up to the political struggles around plant genetic resources in the twentieth century (see also Kloppenburg 1988).[5] In the German-language academic context, Michael Flitner (1995) has presented a similarly comprehensive study of plant genetics, biopolitics, and agricultural modernisation in Germany and in the international agro-political context as they emerged in the twentieth century, but only goes back as far as the late nineteenth century (see also Flitner/Görg/Heins 1998 as well as English: Flitner/Heins 2002; Flitner 2003). Tiago Saraiva (2013) makes an intriguing contribution to this historical-political research by discussing the history of European integration in the aftermath of World War II as a history of transforming crop diversity into a technoscientific "cosmopolitan commons" through coordinated genebanks.

5 The second edition, published in 2004, contains an additional chapter on the dis/continuities of these developments in the early twenty-first century.

He exposes the continuities of this mode of imperialism with "how Nazi and Soviet continental ambitions depended on controlling plant genetic resources" (Saraiva 2013: 186).[6]

Other historians complement these broad historical-political accounts of the role of seed banks and seed conservation for the emergence and consolidation of nation-states and empires with analyses that focus on the underlying imaginaries and historically changing discourses around seed diversity loss and conservation. In an article unfolding a "semantic history of the 'resource' idiom", Christophe Bonneuil (2019: 1) traces "the emergence of a genocentric and resource-centric ontology of the planet's biological diversity" since the seventeenth century. This paper provides a deep historical background to an earlier publication co-authored with Marianna Fenzi that discusses how the politicisation of genetic erosion and seed banking in the second half of the twentieth century gave rise to the biodiversity paradigm, which rebranded *ex situ* conservation efforts without substantially challenging the resourcist view of biological diversity (Fenzi/Bonneuil 2016; see also section 1.4). Sara Peres (2016) contributes to this discourse-analytical historical contextualisation of the emergence of *ex situ* conservation by investigating scientists' imaginaries of seed banking as a solution to genetic erosion. She argues that the idea that frozen seed archives make "the evolutionary history of crops" (Peres 2016: 96) accessible as a future (re)source of knowledge and action is a "memory practice" (Peres 2016: 103, quoting Bowker 2005) that organises the past based on anticipating forms of use in the future. In a subsequent publication, Peres (2019) describes this performative mode of anticipatory future-making through the freezing of seed diversity as an enactment of "cryopower", which refers to the control of biological and associated socio-historical processes through the use of cold temperatures (see also Friedrich 2017; Kowal/Radin 2017).

An interest in the ontological performativity of cultural narratives related to agrobiodiversity loss and conservation also informs the historical research on *ex situ* agrobiodiversity conservation conducted by Helen A. Curry. Her early research focuses on the role of the Rockefeller Foundation in the Green Revolution and in the subsequent establishment of large-scale and long-term crop diversity collections (Curry 2017b), as well as on twentieth century plant genetic

6 According to Curry (2022a: 5), "German, Soviet, and US institutions developed especially ambitious programs of collecting and disseminating crop diversity [in the early twentieth century; FV]" – a history she unfolds in the first and second chapter of the book quoted here.

resource conservation in other contexts (Curry 2017a, 2019a, 2019b). Building on this research, a recent article of particular interest to this study scrutinises the historical context in which the logic of backup emerged and superseded the logic of banking as the dominant conception of future seed and food security (Curry 2022b). In a book synthesising her research, Curry unfolds a history of agrobiodiversity endangerment and conservation through the story of maize cultivation in the United States and Mexico (Curry 2022a). She is critical of the technoscientific overdetermination of the international discourse on agrobiodiversity loss and conservation, and shifts the focus from stories about the nature, loss, and conservation of plants in general to stories about farmers, communities, and ways of life connected to the existence, endangerment, and disappearance of specific plants. The aim of this endeavour is to "tell a history of ideas about people and the consequences of those ideas for the seeds we save, the crops we grow, and ultimately the foods we eat" (Curry 2022a: 6–7).

What all these historical accounts of the role and performativity of seed banks and *ex situ* agrobiodiversity conservation have in common is that they show how ecological and societal realities are emergent and deeply entangled. Speaking with Donna Haraway (2003, 2008), they all – although in various, sometimes more, sometimes less explicit ways – attend to the ways in which history is a process of naturalcultural becoming and doing.[7] The question of how naturalcultural becoming is enacted guides a last body of research I wish to mention here.

Research on Seed Conservation as More-Than-Human Care

A small number of cultural studies scholars have recently focused empirical attention on the labour and specific practices of seed conservation. Building on recent discussions around relations of care in technoscientific and more-than-human contexts (e.g. Hartigan 2017; Puig de la Bellacasa 2017), Xan S. Chacko (2019a, 2022a) and Marleen Boschen (2022) analyse the preparation of seeds for medium- and long-term conservation in *ex situ* environments as a work of

7 The notion of "naturecultures" encapsulates the inextricable relatedness of nature and culture and the relational ontology that characterises Haraway's thinking. For Haraway, being is always becoming with others in "co-constitutive relationships in which none of the partners pre-exist their relating, and the relating is never done once and for all. Historical specificity and contingent mutability rule all the way down, into nature and culture, into naturecultures." (Haraway 2003: 12)

care for naturalcultural futures. Aware of the bio- and cryopolitical entanglements and colonial continuities of global-scale *ex situ* agrobiodiversity conservation endeavours in particular, Chacko engages with the "creative agencies" (2019a: 98) of the scientists "whose labor and decisions will shape that future natureculture" (2019a: 105). Similarly, Boschen questions what she calls the "anthropocentric image of cryopolitical control" (2022: 20) by exploring the radically transformative if precarious potential of seed banks to become means of post-anthropocentric, anti-colonial, and community-based social change. Foregrounding the links between such specific practices, processes, and stories of biodiversity conservation, on the one hand, and conceptions of species, ecosystems, and the (future) earth, on the other hand, Anna-Katharina Laboissière explores how "biodiversity repositories [are] an active intervention into the shaping of natures both inside and outside [the seed bank]" (2019: 65). She therefore proposes to understand seed banks as "world-making devices" (2019: 71–73). However, Laboissière emphasises, as does my study, that neither the practice of conserving plant genetic diversity in *ex situ* repositories nor those repositories themselves determine which of the multiple "possible world-making projects banked along with the collection of seeds" (Laboissière 2019: 76) is eventually going to be realised.

Studying Agrobiodiversity Conservation as a Mode of World-Making

The analysis of the Svalbard Global Seed Vault and its agrobiodiversity conservation efforts that I develop in this book joins the ranks of these approaches that conceive seed banking as a mode of world-making. In contrast to the aforementioned empirical studies of other seed banks, however, my focus lies not on practices of world-making through seed banking – at least not the kind of hands-on conservation practices that handle seeds pre- and post-cold storage. Conducting an empirical investigation of the Seed Vault means studying a seed bank that differs considerably from other seed banks insofar as it is a repository for the latter to securely store duplicates of their collections for the long term, while leaving the preparation and postprocessing of the seeds to the depositor. This means studying an *ex situ* conservation environment where "conservation practices" are not only restricted to depositing sealed boxes of packaged seeds inside storage rooms and monitoring the functioning of the facility, but one that also restricts access to the facility for reasons of security.[8] My research experience illustrates that while this closes the door (in a very literal sense) to "what happens within the apparatus of these repositories" (Laboissière 2019: 65), at the same time, it opens a window to the worlds the Seed Vault is situated in. The latter refers, on the one hand, to the globally extending *ex situ* conservation environment the Seed Vault is part of and backup for. On the other hand, it also refers to the plurality of worlds within what the dominant modern cosmology conceives as *the* (one) world – a claim that I unfold throughout the chapters that follow. Following Haraway's emphatic argument that "[it] matters what worlds world worlds" (2016: 35), the focus of my study is on the worlds participating in the world-making endeavour that the Seed Vault embodies as well as the worlds it makes, makes possible, and could make possible.

Almost three decades ago, Haraway first argued that "[n]othing comes without its world, so trying to know those worlds is crucial" (1997: 37). At that time, this was an exercise in breaking down the scientistic "wall between the political and the technical" (ibid.) and developing a type of knowledge about technoscientific projects that claims objectivity based not on stripping these projects from their worlds but rather on acknowledging those worlds and their epistemological and ontological politics. Following this approach to the study of technoscientific projects, I concentrate my attention on the worlds involved in the Svalbard Global Seed Vault and the worlds that come with the seeds collected inside the vault. This analytical approach is based on the conviction that the Seed Vault is a conservation facility that cannot be understood in itself, because it is a nodal point of an agrobiodiversity conservation environment that extends across space and time. This means that the Seed Vault cannot be understood without an appreciation of its relations to the many national, subnational, and transnational genebanks involved as depositors, each of which comes with its world, meaning a specific entanglement in and emergence from the history of agrobiodiversity conservation unfolded in the previous chapter, and situated relations with the more-than-human ecologies

8 Accounts of the Seed Vault based on visits made before October 2016, such as Heatherington and Perley's "Fieldnotes from Svalbard" (2017), show that access to the Seed Vault was granted in the first few years after it opened. This changed after a water intrusion in October 2016 and subsequent technical reconstruction of the facility, as I elaborate in chapter 3.1.

its conservation efforts co-produce.[9] Paying attention to these worlds that come with the (human and nonhuman) agencies involved in the Seed Vault shows that it is also a nodal point of various discourses, imaginaries, and doings each playing their part in the worlding the Seed Vault is engaged in.

With the proposed mode of investigation – exploring the Seed Vault not in itself but through its relational entanglement in multiple worlds and worldings – comes an understanding of seed banking that I share with some of the abovementioned authors, namely as an indeterminate mode of world-making. In other words, I argue that the Seed Vault's worlding is not determined by the fact that it is a technoscientific project but depends on the worlds it is used to supply and sustain. In this indeterminacy lies a potential for seed banking to be more than a "technological fix" for problems that are actually political, as critics often characterise technoscientific developments in the modern agro-industrial system (e.g. Ribeiro/Shand 2008; Fowler/Mooney 1990: 59; Subramaniam 2024: 165). Transformative potential is usually ascribed to local farmers' and community seed banks, whereas inter- and transnational genebanking projects such as the Svalbard Global Seed Vault are typically associated with techno-salvationism (see also Breen 2015). In contrast to such a perspective, I argue that this juxtaposition harbours the danger of a simplistic romanticisation of peasant farming, on the one hand, and aversion to technoscience, on the other hand. This dualistic understanding suggests that practices, technologies, and their performative effects are inseparably linked with the worlds from which they emerge. While this might be true for some technoscientific projects, I argue that the Seed Vault harbours a potential to be an important tool of social-ecological transformation in light of agrobiodiversity loss and destruction precisely because it assembles not only a diversity of seeds but also a diversity of worlds.

Certainly, this does not mean dismissing the critical analyses summarised above and succumbing to an interpretation of the Seed Vault that is simplistic and romanticising in a different way, namely by seeing it as an innocent

9 The *ecological* entanglements and relationalities of the seeds conserved in the Seed Vault lie outside the scope of my research for two main reasons. First, in the Seed Vault, which is the empirical starting point and heart of my inquiry into the relationalities within which it exists, seeds are as far away from their ecological entanglements as they get in the process of conservation. Second, the ethnographic research I had planned to conduct in seed conservation environments handling seeds within their ecological entanglements was thwarted by the COVID-19 pandemic, as elaborated in the following chapter.

technoscientific means of multilateral future-making. On the contrary, I follow the existing research in attending to the ways in which the Seed Vault is a technoscientific promise of security in the face of a multiplicity of social-ecological threats (3.2), of extending a threatened present (4.1), and of salvation in the face of eco-apocalyptic expectations (5.1). As others have done for other seed banks, I discuss continuities of extractivist colonial histories that come to matter in the Seed Vault, albeit by tracing them not in its (rather short) institutional history so much as in the history of the more-than-human relations materialising in the Seed Vault (3.1). In addition, I complement these analyses by also attending to unexpected worlds and worldings the Seed Vault assembles and the transformative potential which arises from encounters and collaborations of different worlds that emerge in the context of the Seed Vault's efforts to assemble the world's agrobiodiversity in Svalbard (3.3 and 4.2).

2.2 Research Design

The analysis I develop throughout this book conceives of the Svalbard Global Seed Vault not merely as a technoscientific conservation facility assembling a diversity of seeds, but as an arena assembling a diversity of worlds and modes of worlding. My main interest is in how these world(ing)s shape – each in their own way as well as collectively in their historical and ongoing interplay – how agrobiodiversity, as well as its loss and conservation, come to matter in various ways, and what promises and problems these entail. The empirical foundation of my inquiry comprises ethnographic field research, qualitative expert interviews, and document and video analysis. The empirical-analytical process was conducted following the methodology of Situational Analysis (Clarke 2005; Clarke/Friese/Washburn 2018). In the following pages, I delineate the methodological principles as well as the empirical process and data underlying the research presented in the chapters that follow.

The Methodology of Situational Analysis

The empirical and analytical-interpretive process was guided by the qualitative-interpretive methodology of Situational Analysis. First developed by Adele E. Clarke (2005) and later in cooperation with Carrie Friese and Rachel Washburn (Clarke/Friese/Washburn 2015c; 2018; 2022), Situational Analysis is an extension of Grounded Theory in the tradition of Anselm Strauss (1987;

Strauss/Corbin 1990). Whereas the latter is rooted in symbolic interactionism and pragmatism (and inseparably so, according to Clarke), in developing Situational Analysis, Clarke aspired to "push grounded theory around the postmodern turn" (Clarke 2015: 91; 2005) and the interpretive turn (Clarke/Friese/Washburn 2018). While Situational Analysis shares some of the theoretical and methodological assumptions of Straussian Grounded Theory as well as of the post-Straussian Constructivist Grounded Theory of Kathy Charmaz (2006, 2014), it also embraces theoretical approaches such as Michel Foucault's theory of discourses and their "conditions of possibility", the attention to the more-than-human advanced by STS, and theorisations of the situatedness of knowledge and research such as formulated by C. Wright Mills and Haraway (see Clarke/Friese/Washburn 2015a: 12–13, 43–47). These influences converge in the conceptualisation of the *situation of inquiry*, which is the fundamental unit of any Situational Analysis – rather than individual subjects or collective actors, for instance (Clarke/Friese/Washburn 2015a: 12). This makes it possible to bring four facets into focus that often remain peripheral if not impalpable in traditional Grounded Theory and that are crucial to my inquiry into the world(ing)s of agrobiodiversity loss and conservation. Situational Analysis shifts the focus of analysis (a) from social action to *social ecologies* and the relations that make them and (b) from knowing subjects to *discourses*. It further extends the analytical focus to (c) *nonhuman agencies* as well as (d) *implicated actors* and actants, i.e. those that are silenced or physically excluded in a situation of inquiry.

The first shift is articulated through an extension of Strauss's social worlds/arenas theory and informs the particular analytical approach to the Seed Vault and the world(s) of agrobiodiversity conservation this book unfolds. In Situational Analysis, social wholes are not conceived as bounded groups (e.g. geographically) or as the sum of their parts (e.g. individuals). Rather, they are "ecolog[ies] of relations" (Clarke/Friese/Washburn 2015a: 45, referencing Star 1995) meaning "emergent and fluid entities" (Clarke/Friese/Washburn 2018: 148) held together by shared interests or concerns addressed through shared discourses, technologies, and collective action. *Social worlds*, then, denote ecologies of relations that comprise various individual and collective actors, who, in turn, can be part of various social worlds. *Arenas*, in contrast, are areas of sustained interest that assemble different social worlds whose actions and discursive engagements co-construct the respective arena and its participants. Insofar as the perspectives, concerns, and interests of the actors involved diverge, arenas are sites of dispute and thus serve to analyse the workings of power relations. Accordingly, the empirical and analytical focus lies less on individual knowing subjects than on discursive formations and social realities as they come to matter in the arenas under investigation (Clarke 2015: 88–91; Clarke/Friese/Washburn 2015b: 174–176).

This conceptual matrix informs the approach to the Seed Vault I propose, namely seeing it not primarily as a particular technology or set of technoscientific practices that is part of a particular social world, but as an arena assembling a diversity of (social) worlds that come with a diversity of technologies, practices, actors, concerns, discourses, disputes, etcetera. This makes it possible to foreground ambiguities, unanticipated partners and relations, and spaces of transformative possibility in the biopolitics of the Svalbard Global Seed Vault and thus expand the scope of previous analyses.

Identifying cracks in "master discourses" (Clarke/Friese/Washburn 2015a: 21) and promoting "epistemic diversity" (ibid.) is one of the explicit goals of Situational Analysis. In this sense, it particularly "lends itself to critical, feminist, anti-racist, and related projects in several important and very intentional ways" (Clarke/Friese/Washburn 2015a: 20). Developed out of an effort to abandon residual tendencies of the methodology of Grounded Theory towards othering and representation and to overcome oversimplifications that aim to purify the research outcome (Clarke 2005: 11–18), Situational Analysis cultivates attention to "difference(s), power, contingency, and multiplicity" (2005: 11). In an attempt to avoid unwittingly reproducing tacit "hierarchies of importance" (Clarke/Friese/Washburn 2015a: 21) through the design of a research project, the methodology of Situational Analysis is designed to map "*all* elements, *all* positions, *all* voices" (ibid., emph. in orig.) that are part of an arena or a situation of inquiry.

Firstly, this means attending to *nonhuman actors and elements* – both discursively and materially – based on the premise that their presence or absence in a situation matters in the sense that they co-constitute a situation and the conditions of possibility for action and interaction it entails (ibid.; Clarke 2015: 91–93). Crucially, this involves not only organic beings such as (in the case of this research project, most obviously) seeds, but also infrastructures, technical entities, symbols, and suchlike. Further, comprehensive mapping involves lending visibility to *implicated actors*, that is, human and nonhuman agencies who are either physically present but silent or silenced or only discursively made present (in one way or another) in the situation of inquiry. Finally, it also includes a critical reflexion of the *researcher's own situatedness* in terms of both their implication in the situation of inquiry and their more general social positionality. The double aim of this is to increase the researcher's accountability, on the one hand, and to reposition the researcher as a situated expert for a particular and partial perspective on the situation of inquiry and hence a learner vis-à-vis the subjects of research, on the other.

These methodological premises have two main methodical implications that have guided the research process underlying my study. The first is that Situational Analysis is a "*multi-site or multi-modal* research [approach]" (Clarke/Friese/Washburn 2015a: 16, emph. added). Unlike in multi-sited ethnography (Marcus 1995, 1998), multi-sitedness, here, does not necessarily imply that research is conducted throughout multiple geographical sites. Instead, it refers to a multiplicity of data sources (which can correspond to different geographical sites of research), levels of research, and perspectives. Situational Analysis encourages adducing multiple types of perspectives and data including interview, ethnographic, and other discursive material as in the empirical basis of this study, based on the assumption that a situation is a messy and complex web of various elements and relations entangled in discourses and structures that exceed and at the same time enable the situation of inquiry. The goal is to explore the relations between the various elements present in and constituting a situation, such as "human and nonhuman actors and actants, institutions and organizations, discourses and symbols, political economies and sociocultural landscapes, and the like" (Clarke/Friese/Washburn 2015b: 172). This makes it possible to reveal the conditions of possibility of a situation, which Situational Analysis, unlike traditional Grounded Theory, assumes to be *in* the situation rather than any sort of external context of the situation: "Regardless of whether some actors might construe them as local or global, internal or external, close-in or far away, or whatever, the fundamental question is: '*How do these conditions appear – make themselves felt as consequential – as integral parts of the empirical situation under examination?*'" (Clarke 2015: 98, emph. in orig.)

To bring the various elements, relations, and conditions that constitute a situation of inquiry into view and analyse them, the second methodical proposition of Situational Analysis is *mapping* them. The method and process of mapping is the heart of any Situational Analysis. It differs significantly from the coding process of Grounded Theory insofar as the latter operates through a consecutive analytical process that condenses codes into categories, categories into concepts, and concepts into an empirically grounded theory. The mapping process in Situational Analysis, in contrast, aims to develop a multi-faceted and broad understanding of the situation of inquiry and the relevant elements and actors, relations, and discourses. Unlike traditional

Grounded Theory, then, it is not a linear, inductive method but a helical or iterative and generatively messy, *abductive* process of reflexively moving "back and forth between the nitty-gritty specificities of *empirical* data and more *abstract conceptual* ways of thinking about them" (Clarke/Friese/Washburn 2018: 28, emph. in orig.). By integrating rather than separating empirical research, analysis, and theorising, Situational Analysis grounds decisions about what to examine more closely in the situation. It is thus a cognitive as well as an experiential process of curiously and carefully getting to know an object or field of research, making room for surprises and generative doubt, embracing rather than prematurely simplifying heterogeneities, ambiguities, and incoherencies, and thus arriving at situated, accountable, and theoretically grounded useful explanations (see Clarke/Friese/Washburn 2018: 28–32).

The Empirical-Analytical Process

The findings presented in this book are the result of such a curiously open and abductive research process inspired for one by Situational Analysis but also borne by an ethnographic ethos that does not predetermine the paths and goals of the research in advance but rather allows them to emerge throughout the research process. Such an open approach is particularly conducive to cultivating attentiveness to the "polyphony" (Tsing 2015: 23) of heterogeneous yet entangled world-making efforts in a given research situation. The process of compiling empirical data can be roughly divided into a first, explorative phase of research in 2019, a second, immersive phase ranging from 2020 to early 2021, and a third, conversational phase after that. While the first two phases comprised all modes of research, the core of the last phase were two interviews. One is my interview with Cary Fowler, the founder of the Seed Vault, whom I approached not to hear his well-documented story in oral form so much as to discuss my insights and questions with him. The other is an interview with an engineer from Statsbygg, the company responsible for Norwegian government estates and hence the Svalbard Global Seed vault, the opportunity for which presented itself spontaneously in March 2023.

Document Analysis

At the beginning of the research process, I entered the field through an extensive engagement with online information material, news reports, documentaries, and prior academic research about the Svalbard Global Seed Vault, as well as historical documents about agrobiodiversity loss and conservation such as books, reports, and treaties. Rather than conducting a systematic document analysis, I collected this data following a snowball principle to map the situation of inquiry and all the relevant actors, entities, and discourses as comprehensively as possible, in order to allow for a grounded decision on what to focus on more closely. Throughout the later phases of the research process, I collected text and film documents that gave me deeper insights than those gained through other sources of empirical data or field access that was otherwise obstructed.

Ethnographic Research

The ethnographic research conducted for this study comprised two field trips to Svalbard, one of ten days in June 2019 and one of seven days in February 2020, as well as a one-day visit to the Leibniz Institute of Plant Genetics and Crop Plant Research (IPK) in Gatersleben, Germany, in December 2019. Whereas the purpose of the two research trips in 2019 was mainly exploratory, the occasion of the second trip to Svalbard was an international seed summit and anniversary event at the Seed Vault. All three ethnographic episodes provided me with valuable insights into the workings, networks, and rationalities of the *ex situ* world of conservation, allowed me to explore the situatedness of the Svalbard Global Seed Vault within this world as well as in its home, Svalbard, and enabled me to make important contacts, acquire interview partners, and engage in further informal exchange with members of the *ex situ* world.

Guiding my ethnographic research was a special attentiveness to the human and more-than-human relations constituting the world of agrobiodiversity conservation, inspired by relational approaches to ethnography in which encounters and processes of becoming are of particular interest (Desmond 2014; Faier/Rofel 2014; van Dooren/Kirksey/Münster 2016). Since the circumstances did not allow for a multispecies ethnography (Kirksey/Helmreich 2010; Kirksey/Schuetze/Helmreich 2014) attending to more-than-human relations as they come to matter in seed conservation facilities, which I had originally planned to carry out, more-than-human relationality became a "sensitising concept" (Clarke/Friese/Washburn 2018: 54, quoting Blumer 1969: 147–148) more than a core focus of my research.

Generally, ethnographic research was complicated by the high costs of travelling to Svalbard as well as by the scattering of the relevant actors connected to the Seed Vault across the whole world. Additionally, further ethnographic research I had planned to conduct (for instance, at the IPK) was thwarted by the national and international travel restrictions imposed in response to the COVID-19 pandemic shortly after my second journey to Svalbard. When the restrictions were lifted, I had adjusted the research plan in a way that additional ethnographic research would have complicated more than benefitted the process. The book at hand, therefore, is not an ethnography in the narrower sense, but rather comprises elements of ethnographic research and data. These approaches inform my basic understanding of the field under investigation. They are also included in the text in the form of vignettes, which function as testimonies of my ethnographic experiences synthesised from field notes and my ethnographic diary, with the aim of taking the reader into the situation discussed in the respective chapter (on vignettes, see Gullion 2016: 89–90). These text passages are italicised. Where they contain direct speech by informants I encountered in my ethnographic research, these passages paraphrase conversations reconstructed from field notes. Hence, the respective informants are anonymised and not quoted in the same way as the interviewed experts.

Expert interviews

The third empirical pillar of this study consists of fourteen semi-structured qualitative expert interviews conducted between May 2019 and March 2021, with one exception in March 2023. The interviewees are members of the institutions directly involved with the Seed Vault, other relevant institutions and organisations involved in *ex situ* conservation practices and politics, and representatives of the non-governmental and civil society sector (see table 1). While some essential informants were easily identifiable from the outset, in other cases it only became clear during the research process who might be an interesting interviewee. Following the abductive logic of Situational Analysis, I therefore selected my interview partners in an open and continuously evolving process of theoretical and empirical engagement with the world of agrobiodiversity conservation.

Table 1: List of Expert Interviews (the author, 2025)

No.	Date	Name	Institution	Profession
01	28 May 2019	Anonymous[i]	Crop Trust (Bonn, Germany)	Communications
02	28 May 2019	Hannes Dempewolf	Crop Trust (Bonn, Germany)	Scientist, background: (agricultural) botany
03	6 Jun 2019	Anonymous[i,ii]	Norwegian Polar Institute (Longyearbyen, Svalbard)	Scientist, background: climate science
04	6 Jun 2019	Åsmund Asdal[ii]	NordGen (Alnarp, Sweden)	Seed Vault Coordinator, background: horticulture, scientific management
05	6 Mar 2020	Anonymous	FAO (Rome, Italy)	Scientist, background: conservation genetics
06	12 Mar 2020	Godfrey Mwila	Various	Founder of the national genebank of Zambia; representative of the African Group in the negotiations of the Seed Treaty; collaborations with Bioversity International and the Crop Trust
07	26 Mar 2020	Kent Nnadozie	Governing Body of the ITPGRFA, UN/FAO (Rome, Italy)	Secretary, background: environmental law
08	1 May 2020	Henk Hobbelink	GRAIN (Barcelona, Spain)	Agronomist, co-founder of GRAIN
09	26 Mar 2020	Pat R. Mooney	ETC Group (formerly RAFI) (Montréal, Canada)	Agrobiodiversity expert, co-founder of the ETC Group
10	26 Mar 2020	Anonymous	AHTEG, Seed Treaty, UN/FAO (Rome, Italy)	CSO representative

No.	Date	Name	Institution	Profession
11	5 Nov 2020	Alejandro Argumedo	Potato Park / Asociación ANDES (Cusco, Peru)	Indigenous activist, inhabitant of the Potato Park, co-founder of the Asociación ANDES
12	20 Jan 2021	Anonymous	Bioversity International (formerly IBPGR/IPGRI) (Rome, Italy)	Scientist, background: plant science and conservation genetics
13	19/20 Mar 2021	Cary Fowler	Various	Founder of the Svalbard Global Seed Vault (see ch. 1 for further positions)
14	26 Mar 2023	Anonymous[ii]	Statsbygg (Oslo, Norway)	Engineer

[i] interview partners not quoted in the text
[ii] interviews conducted in person

The expert interview as the method of choice is rooted in the understanding of expert knowledge within the qualitative research paradigm. It is targeted as a specific type of knowledge offering specific kinds of insights and access into a field of research rather than as a source of objective knowledge about a field being studied. Alexander Bogner and Wolfgang Menz (2009: 46–48) distinguish three types of expert interviews that serve different purposes in the research process, all of which informed the expert interviews conducted as part of this study. First, *exploratory interviews* allow the researcher to gain an overview of a field of research and possibly access to further sources, as well as to structure and elaborate their interest. This type of interview targets experts as sources of contextual knowledge about the field being studied and as "crystallization points" (Bogner/Menz/Littig 2009: 2). Second, *systematising interviews* aim for deeper and more systematic access to the technical, processual, and experiential knowledge of the expert who takes on the role of a guide. Third, *theory-generating interviews* communicatively target and serve to analytically reconstruct the interpretive and "subjective dimension of expert knowledge" (Bogner/Menz 2009: 48). The typical method for expert interviews, which also structured the interviews conducted for this study, is the semi-structured interview. This is based on a prepared compendium of open questions guiding the interview while at the same time leaving room for the interviewee to expand on issues and perspectives beyond the questions explicitly posed (Flick 2014: 197).

While the first couple of interviews, which I conducted in 2019, had an exploratory character, they were pre-structured by the initial document analysis. The interviews that followed were by and large systematising expert interviews, however, with open questions aiming to prompt narrative passages expanding on the interviewees' interpretive and subjective knowledge. Except for three in-person interviews, all others were conducted via video call, in some cases due to geographical distance, in others because of travel restrictions due to the COVID-19 pandemic. The questions and themes guiding the semi-structured expert interviews were generally orientated in the same direction, but tailored to the respective interview partner as well as continuously developed in response to the data already collected. The interview guides, therefore, differ in content although they resemble each other in structure. By and large, all interview partners were interrogated about their own background and involvement in the world of agrobiodiversity conservation, about key actors, developments, challenges, and conflicts in the history and present state of agrobiodiversity conservation, as well as about the Svalbard Global Seed Vault.

Throughout the book, readers will find that some interviewees are quoted by name whereas others are not. This has to do with the unique role or position some of the interviewees have in the field, in which case an anonymisation would either be a major loss of information or hardly feasible at all. Those interviewees who remain anonymous are either members of important but large institutions which they do not represent as a whole, or do not wish to be quoted by name. Two interview partners are not directly quoted at all throughout the book and remain background informants. All interview passages that appear throughout the text are quoted from written transcripts of the interviews. At the request of some of my interview partners and on the grounds that the analysis focuses on content rather than being depth-hermeneutical, the quoted passages are copyedited so that they no longer contain pauses, um's, stuttering, grammatical errors, and suchlike. Since I have been careful not to distort any of the content of the interview passages, they are still clearly recognisable as spoken language. Citations of interview transcripts throughout the text indicate the speaker (by name and/or job description), the institution they speak for, the date of the interview, as well as the line numbers of the respective passage in the transcript.

The Analytical Mapping Process

Throughout the research process, I visualised, organised, analysed, and reflected the empirical data and research findings through the creation of the various types of maps developed by Clarke and colleagues to digest the empirical data, get "a grasp of the big picture" (Clarke/Friese/Washburn 2018: 150), and come to grounded decisions about which key arenas, worlds, actors, technologies, discourses, etcetera to focus on. First, *situational maps* (Clarke/Friese/Washburn 2018: 127ff.) allowed me to assemble all the elements making up the situation of inquiry – from individual and institutional actors to technological, discursive, and other nonhuman elements and so forth – and trace the relations between them.

Social worlds/arenas maps (Clarke/Friese/Washburn 2018: 147ff.) served to identify the key arenas in the situation of inquiry and the various social worlds they assemble. It was through this type of mapping that the Seed Vault became evident to me as not only a technology or set of technologies but an arena. This, in turn, led me to shift attention from the diversity of seeds assembled inside the vault to the diversity of worlds that come with those seeds or with other material-discursive elements in the arena that is the Seed Vault (see chapter 3.2). Thus crystallised, for instance, what I have come to call the *ex situ* world of conservation (see chapter 4.1) but also the situatedness of the Seed Vault within a world and mode of world-making that has a much deeper history than the Seed Vault itself (see sections 3.1 and 5.1). I use the term "world" instead of "social world" in order to avoid the anthropocentric tone and organisational bias of the latter, and – as unfolded in the introduction to this book – in reference to academic discourses about the plurality and performativity of worlds (see esp. Law 2015; Haraway 2016; de la Cadena/Blaser 2018a).

Finally, through *positional maps* (Clarke/Friese/Washburn 2018: 165ff.) that concentrate on how "[s]ocial worlds are 'universes of discourse'" (Clarke/Friese/Washburn 2018: 16, quoting Strauss 1978), I was able to identify key discursive issues in the arena that is the Seed Vault as well as the range of discursive positions on these issues including less powerful or marginalised ones. Positional maps are explicitly designed to help researchers avoid reproducing the dominance of certain voices and positions as well as the epistemic power relations within the situation of inquiry. The result of my work with these maps is reflected in the structure of chapters 4 and 5, which contrast divergent positions on what it means to *conserve* agrobiodiversity (see chapter 4) and how to deal with the *loss* of agrobiodiversity (see chapter 5), and illustrate how they are rooted in and enact different worlds and modes of knowing and being in the world.

3. Exploring the Naturalcultural History and Politics of Svalbard and the Global Seed Vault

In the 16 years since the Svalbard Global Seed Vault first opened its doors to the world's seeds and genebanks as well as to the public, it has attracted a lot of international attention. Journalists from all around the world have disseminated its story, as have a number of social scientists and humanities researchers. Many of these stories have been told from afar, based on sources of knowledge accessible from afar, because its remoteness and the high cost of a journey to Svalbard make it difficult to visit the Seed Vault. In recent years, access to the Seed Vault has become even more difficult since its doors were closed to the public in the wake of a water intrusion incident in 2016 and the subsequent reconstruction of the facility. On both occasions when I travelled to Svalbard for field research, in June 2019 and in February 2020, the security door separating the interior of the vault world from the outside world remained closed for me. It was not least this restriction of access to the interior world of the Seed Vault that prompted me to turn my attention outward – to the world(s) it is situated in. While this choice was imposed at first, it opened insightful, important, and hitherto underexplored paths towards the many worlds the Seed Vault is entangled in. It prompted me to consider and investigate the Seed Vault not only as a technoscientific conservation facility for seeds from all over the world, but also as a nodal point of the many worlds that come with the seeds and other actors the Seed Vault assembles and hence a window to the world(s) of agrobiodiversity conservation.

As elaborated in the introduction to this book, this approach to studying the Seed Vault and agrobiodiversity conservation follows the research tradition of feminist technoscience studies, which – together with postcolonial and other critical (techno)science studies – place the worldly situatedness and performativity of technoscientific projects, practices, and knowledges at the centre of attention. Investigating the worlds that come with technoscientific objects, knowledges, and practices, as Donna Haraway has prominently proposed, is an effort to tear down "the wall between the political and the technical" (Haraway 1997: 37). It means challenging the isolation of the sciences from critical reflexivity and recognising the ways in which the technical is imbued with the political. Paying attention to the worlds that come with objects and subjects of research, Haraway argues, is a contextualising exercise that is indispensable for any critical inquiry insofar as it highlights how technoscientific knowledge- and world-making practices and processes are always located in specific worlds, embedded in power relations, and never impartial, disengaged, or innocent (see also Haraway 1988).

The present chapter introduces what it means to consider the Svalbard Global Seed Vault through the worlds it comes with and that come with the seeds it assembles. The first section focuses on the world the Seed Vault is situated in in its home, Svalbard. While Svalbard typically matters in stories about the Seed Vault in terms of the geographical, geological, and climatic environment it provides, my analysis focuses on the history of the archipelago, exploring how it comes to matter in the story of the Seed Vault. The second section traces the story of the Seed Vault through the eyes of the various actors in the world of agrobiodiversity conservation who are involved in it and to whose work it matters. It is through these accounts that I develop the analytical perspective I propose, namely to conceive the Seed Vault not merely as a technoscientific facility and actor in the world of agrobiodiversity conservation, but also as a political arena assembling a diversity of worlds and modes of world-making in the interplay of which the Seed Vault and its conservation efforts come to matter.

3.1 Svalbard: "The Last Frontier"? Dis/continuities of Frontierism in *ex situ* Conservation

> To understand the Seed Vault, you first need a feeling for its home, Svalbard. (Fowler 2016: 25)

At the beginning of the first chapter of *Seeds on Ice: Svalbard and the Global Seed Vault* (Fowler 2016), Cary Fowler takes his readers into the extraordinary environment that Svalbard provides to the Seed Vault. This environment is shaped by an icy mountainous landscape, endless darkness in the long Arctic winter and all-day sunlight in summer, as well as a sparse flora and unique fauna adapted to the Arctic climate and milieu. Stories of humbling experiences the challenging environment of Svalbard has afforded Fowler himself and other men throughout history imbue the ensuing overview of the little more than four centuries of human history on the archipelago. A closer engagement with the history of human activity on and interest in the archipelago shows that Svalbard matters to the story of the Seed Vault in more ways than due to its remoteness, geological stability, and cold temperatures, which are the usual aspects mentioned when the archipelago is presented as a suitable home for the Seed Vault. This chapter shows that Svalbard also gives the Seed Vault a more-than-*non*human home.[1]

It is not immediately obvious what the history of Svalbard might have to do with the Seed Vault. The facility has only been part of Longyearbyen's townscape for less than two decades, none of the relevant actors behind the Seed Vault are based in Svalbard, and with no permanent staff the facility is an appendix to rather than a part of the town and the local community. If history is understood in institutional terms, the Seed Vault has no deep historical roots in Svalbard. However, history matters in other ways, too, which might be less evident but are no less meaningful and consequential. The history embedding the Seed Vault that I trace in this chapter is not the institutional history that has given rise to the Seed Vault as a backup seed conservation facility, which I have unfolded in chapter 1. Rather, what I am interested in here is the history of human relations to the natural world as it comes to matter in the history of Svalbard, including the Seed Vault as its most recent materialisation. This "naturalcultural history" (Haraway 2016: 118) at once exceeds and manifests in the history of Svalbard, making it possible to uncover the kindred relations between the Seed Vault and its home, Svalbard.

---

1 Around the turn of the millennium, the notion of the "more-than-human" became a popular indicator of the fact that human lives and life worlds are deeply entangled with more-than-human others and conditions (see Whatmore 2002: 146–167). Although the term aims to encapsulate the relational ontology of the human, it also maintains a distinction between the human and that which is *more* and *other* than human. In this respect, it foregrounds an anthropocentric perspective on relational ontology. The notion of the "more-than-*non*human" is a play on words that aims to subvert this anthropocentric language by highlighting that the way in which the natural world is just as more-than-nonhuman as the human world is more-than-human. I elaborate on the notion in more detail in chapter 4.1.

To trace Svalbard's history, I follow the path laid out by the Svalbard Museum in Longyearbyen. From the first historically verified landing on the archipelago in the course of mid-second-millennium European explorerism through a long history of extractivist resource exploitation over the following centuries, the Svalbard Museum presents the story of what I read as a world made through iterations of frontierism. Drawing on Anna L. Tsing's (2005) notion of the "salvage frontier" as a historically specific contemporary mode of frontierism, I argue that the frontierist world-making trajectory has not come to an end with the growing comprehension of the ecological repercussions of colonial-capitalist modernity in the twenty-first century and the surge in efforts to conserve what is left of the 'old world'. Rather, I show that it is reiterated – even conserved – by the most recent addition to Svalbard's more-than-nonhuman landscape, the Svalbard Global Seed Vault.

The History of Svalbard

The Svalbard Museum sits at the lower end of the centre of Longyearbyen, facing the town to one side and the Isfjorden to the other. It is an asymmetrically built, one-storied, sprawling, dark-coloured building that seems to mimic the mountainscape on the other side of the fjord, which rises above it in the background when viewed from the town centre. The Svalbard Museum was one of the first places in Longyearbyen I was drawn to during my first visit to Svalbard in June 2019. I had arrived two days prior to my arranged meetings with the Seed Vault people in order to have some time to familiarise myself with the place. Upon arrival at the museum, I realised that as part of the Forskningsparken, the Svalbard Research Park, it shares a building with the Arctic outpost of the Norwegian Polar Institute and the local University Centre. Glancing through the simple glass door that separates the latter from the reception hall of the museum, a remark in the work of decolonial philosopher Walter D. Mignolo that I had recently read popped into my head. In *On Decoloniality: Concepts, Analytics, Praxis* (Mignolo/Walsh 2018), Mignolo argues that museums are places of building historical memory as much as they are places of learning about history. Along with universities, he writes, they have historically been "instrumental in building Western civilization's profile and identity, as we know it today" (Mignolo 2018: 199). In doing so, they have exerted and secured what Mignolo calls the "epistemic totalitarianism" (2018: 195) of the European colonial project. I entered the museum sensitised by this reminder that the historical knowledges presented in museums are situated (hi)stories of the events and contexts of which they tell.

As a researcher with only German and English language skills, I found the Svalbard Museum one of the main sources of accessible knowledge about the history of the archipelago, since most written historical sources are in Norwegian or Russian. English-language exceptions further undergirding the following account are Thor B. Arlov's *A Short History of Svalbard* ([1989] 1994) and two short entries in Marc Nuttall's *Encyclopedia of the Arctic* (2005). These accounts largely correspond to the museum's historical narrative, not least because Arlov, a historian at the Norwegian University of Science and Technology in Trondheim, has authored some of the texts on the Svalbard Museum's information boards and website.

Entering the exhibition, visitors embark on a circular path along which they learn about the history of human activity on Svalbard in chronological order. The parqueted path forms an open circuit around an area covered with grey stones on one side and blue ones on the other side, partitioned by and traversable on wooden gangplanks. This area displays the diversity of land- and sea-based nonhuman life on Svalbard, primarily in the form of stuffed Arctic animals accompanied by information signs. Svalbard's fauna comprises reindeer, arctic foxes, birds, and a declining number of polar bears, as well as walruses, seals, and whales in the waters around the archipelago (for more details see Humlum 2005: 1975–1977). With the highest average temperature of 5 degrees Celsius during a rather short summer season of usually no more than one hundred days, the flora is similarly limited and restricted to small wildflowers and short grasses which flourish for no more than a couple of months per year (Fowler 2016: 39–40). In retrospect, the arrangement of the exhibition in the Svalbard Museum seemed to me to re/construct a linear narrative of human-centred cultural history (r)evolving around the biological diversity Svalbard has to offer.

The Svalbard Museum's (hi)story of human presence on Svalbard begins with the first historically verified landing in 1596.[2] With the sea routes south of

2 There is no evidence in historical records that there was any Indigenous human life on Svalbard prior to its discovery by continental seafarers (see Kruse 2012: 61; Fowler 2016: 31). Arlov ([1989] 1994: 12–15; Arlov 2005) outlines a historiographical debate about the possibility of human visits to Svalbard before this historically recognised date of its discovery. This debate is based on persistent claims about the presence of Russian hunters on Svalbard in the fifteenth or sixteenth century, Nordic sailors in the twelfth century, and even Stone Age settlements. At the heart of this controversy, according to Arlov,

the European continent controlled by Spain and Portugal at that time, Dutch and other north European seafarers travelled to northern waters in search of a northerly and independent trade route to the East. The Dutch explorer Willem Barentsz and his crew happened upon the uninhabited archipelago on their third expedition in search of a northeast passage (Arlov [1989] 1994: 9). The search for a sea route through the Arctic continued, but before long this was not the only objective for seafarers from various nations as they sailed northwards. Beginning in the early seventeenth century, a long and multi-faceted history of resource extraction began to deplete the diversity of life on Svalbard and in the waters around the archipelago. This started with the systematic hunting of whales for their blubber, which was exploited as a source of oil to be sold on European markets. While the demand for oil in Europe was growing, formerly abundant whaling regions such as Labrador and Newfoundland were already on the brink of exhaustion (Arlov 2005: 4). In Svalbard, whaling was conducted in the fjords and inshore waters until the mid-seventeenth century. When the whales withdrew to open waters, away from the coastal region that had become a life-threatening environment for them, the hunt continued and was intensified at sea. By the end of the eighteenth century, the whale populations around Svalbard were on the brink of extinction and "[t]he western European whaling adventure in the northern seas had come to an end" (Svalbard Museum n.d. d; see also Arlov [1989] 1994: 16–33).

From the eighteenth until the mid-twentieth century, Svalbard became a popular destination for hunters and trappers going after walrus, seals, and birds as well as arctic foxes for their fur. Considering that hunting and trapping was economically insignificant compared to the former whaling, the narrative of the Svalbard Museum suggests that "the desire to be one's own master and experience adventure […] may have been [among] the main motivations" (Svalbard Museum n.d. a). At the turn of the nineteenth to the twentieth century, a number of airborne North Pole expeditions taking off from Svalbard, first by hot-air balloon and later by airship, drew further international atten-

lies "inconclusive evidence and lack of sources that open up for different interpretations, but also because of particular interests in this contested region" (Arlov 2005: 3; see also Humlum 2005: 1977).

tion to the archipelago.[3] At the same time, intensified efforts to scientifically explore the archipelago including its biological and geological resources generated interest in the mineral resources there (see Arlov [1989] 1994: 42–49). While today's Arctic climate does not allow for much vegetation, what is now Svalbard was once home to "forests, ferns, and dinosaurs, whose tracks and fossils can be found on Svalbard today" (Fowler 2016: 39). An immense stock of black coal lying dormant inside Svalbard's mountainscape is part of the archipelago's unusually well-preserved, sedimented history spanning over 400 million years (Humlum 2005: 1975).

Figure 2: Norway and Svalbard on the Globe (Europe Centred)

Credit: TUBS, license: CC BY-SA 3.0, source: https://commons.wikimedia.org/wiki/File:Norway_on_the_globe_(Svalbard_special)_(Europe_centered).svg from July 18, 2025.

Figure 3: Topographic Map of Svalbard

Credit: Oona Räisänen (Mysid), license: CC BY-SA 4.0, source: https://commons.wikimedia.org/wiki/File:Topographic_map_of_Svalbard.svg from July 18, 2025.

With the discovery of these coal deposits, the second large extractive industry on Svalbard began to take shape. When the development of steam tech-

3 Longyearbyen has its own North Pole Expedition Museum, which exclusively covers air- and waterborne expeditions to the North Pole and displays an impressive number of historical documents, photos, and artifacts. See https://northpolemuseum.com/about (last accessed July 18, 2025).

nology at the end of the nineteenth century made the remote archipelago more easily accessible, "an Arctic coal rush developed" (Kruse 2012: 62). Between 1898 and 1920, before Svalbard became first regulated by international law, businessmen from ten different countries registered more than one hundred private land claims. However, by no means all these coal mining endeavours were successful, for a variety of reasons; geological mapping was insufficiently advanced and the necessary resources were often lacking (see Arlov [1989] 1994: 50–57). With the emerging coal mining business came the first year-round settlements on Svalbard. Some of the settlements located close to large coal deposits soon turned into permanent communities. Five coal mining villages established by companies from Norway, Russia, Sweden, the Netherlands, and the USA ultimately remained operative throughout the twentieth century. In 1906, the American Arctic Coal Company founded Longyear City, named after its major stockholder John Munro Longyear. When the Norwegian Store Norske Spitsbergen Kulkompani took over the coal mines in the area in 1916, the town was renamed Longyearbyen ('byen' is Norwegian for town) (Svalbard Museum n.d. b). Today, Longyearbyen is one of two former coal mining villages with coal mines still in operation and, as the largest of three remaining permanent settlements, the administrative, economic, and cultural centre of Svalbard. The other two settlements are Barentsburg, a Russian coal mining village operative to this day, and Ny-Ålesund, a small Norwegian research station. Although all three permanent settlements are located on Spitsbergen, the Svalbard archipelago's largest island, no road connects them, which means that Ny-Ålesund and Barentsburg are only accessible by boat, helicopter, or snow scooter from Longyearbyen (Fowler 2016: 31, 39).

The first laws and regulations to administer human life and economic activity on the archipelago were passed in the early twentieth century in the wake of the proliferation of the international mining industry on Svalbard. While "Svalbard was regarded a no man's land throughout the 19th century" (Arlov [1989] 1994: 60), the territorial struggles among Svalbard's many occupants who attempted to defend their land claims against trespassers created a need for authorities and regulations, especially with the Arctic environment impeding year-round presence prior to the establishment of permanent settlements (Kruse 2012: 62). A first international attempt to establish a legal regime on Svalbard, on the initiative of Norway, was first obstructed by the USA and Germany and then deferred by the outbreak of World War I. In its aftermath, a treaty to regulate access to the archipelago and its resources was developed as part of the Versailles negotiations. At the Peace Conference in Paris on February 9, 1920, Norway, Sweden, Denmark, the Netherlands, the UK, France, Italy, Japan, and the USA signed the Treaty recognising the sovereignty of Norway over the Archipelago of Spitsbergen (see Royal Ministry of Justice 1988).[4] Norway officially changed the name of the archipelago from Spitsbergen to Svalbard in 1925, and this treaty is now commonly referred to as the Svalbard Treaty.

The Svalbard Treaty recognised "the full and absolute sovereignty of Norway over the Archipelago of Spitsbergen" (Royal Ministry of Justice 1988: Art. 1). In doing so, it abrogated the archipelago's prior status as "*terra nullius*" (Peters 2005: 1982, emph. in orig.). By providing a legal framework to regulate coal mining on Svalbard, a Mining Code Norway covenanted to devise (Royal Ministry of Justice 1988: Art. 8) and passed in 1925, the treaty also suspended conflicts between land claimants. Nevertheless, Svalbard's status under international legislation remains exceptional. Although the archipelago officially belongs to Norway and is regulated by Norwegian authorities as of its ratification by all initial signatories in 1925, the Svalbard Treaty grants equal rights to all citizens and companies from signatory countries regarding entering and residing in Svalbard, acquiring and utilising property, and engaging in land- and sea-based animal as well as mineral resource extraction and commercial activities (Svalbard Museum n.d. c; Peters 2005). Norway is only allowed to curtail these rights if and when they conflict with other provisions, such as the goals of environmental protection stipulated in Article 2 of the Svalbard Treaty and recently refined in the Svalbard Environmental Protection Act (Norwegian Ministry of Climate and Environment 2001).

Regulated by the Mining Code, the coal mining industry on Svalbard flourished for several decades. Beginning in the 1970s, however, it was plunged into a series of crises that eventually ushered in a new phase of human life and economic activity on the archipelago. Norway communised public services, established a university centre, built an airport in Longyearbyen, and relaxed restrictions on tourism; research and tourism became the main pillars of human activity on Svalbard. While the decline of the coal mining industry led to a significant reduction in the population of Barentsburg from about 2,500 in the early 1990s to only 340 inhabitants today, the combined population of

4 There are currently 50 signatories to the Svalbard Treaty. For a complete list see https://lovdata.no/dokument/NL/lov/1920-02-09 (last accessed July 18, 2025). Since it is an open treaty, other countries can join the Svalbard Treaty at any time (Peters 2005: 1982).

Longyearbyen and Ny-Ålesund approximately doubled in the same period to a current number of 2,556 inhabitants.[5] Svalbard's population is characterised by a high fluctuation. According to the last population statistics overview for Svalbard published by Statistics Norway (2017), as of 2016 it was composed of around 46 nationalities, with Norwegians constituting the largest part followed by more than 100 inhabitants from each of Thailand and Sweden and over 30 inhabitants from each of Russia and the Philippines. Svalbard's legislation does not stipulate that immigrants need a visa, residence or work permit, but they must have accommodation, employment and sufficient income to secure their own livelihood. While this incentivises international migration to Svalbard, the absence of political and labour rights as well as social security and inclusion mechanisms creates "a grey zone of informal power structures, unequal living, working and housing conditions, language barrier and perceived low level of social participation" as Zdenka Sokolickova (2022: 293) shows with regard to the Thai and Filipinx diasporas in Longyearbyen.[6] With life on Svalbard being strictly organised around work, there is no social security system and no medical care beyond primary care (Statistics Norway 2017: 10, 23). Severely ill, heavily pregnant, and dying people must relocate to the Norwegian mainland or other countries in order to receive proper care. Sophia Roosth (2016: n.p.) trenchantly summarises the situation: "Svalbard is not, as citizens call it, a 'life cycle community' – no concessions are made for birth and death, and only able-bodied working adults are welcome."

Among Longyearbyen's population, approximately a third works in the administrative, government, or research sector, while two thirds are employed in the declining coal mining business and the growing tourism industry (Fowler 2016: 31–32). This ratio and the development of the local industries also characterise the material townscape. Next to residential housing, there are government and administrative buildings, a higher education and research facility, a library and cultural centre, and a couple of museums, as well as many hotels, restaurants, bars, and souvenir shops complemented by endless tourism

5 For periodically updated population statistics see https://www.ssb.no/en/befsvalbard / (last accessed July 18, 2025, last updated March 31, 2025).

6 An-Magritt Jensen (2009) has identified three main reasons for the migration flow from Thailand to Svalbard: marriage and labour migration, and the search for better living conditions. According to Sokolickova (2022), these results are no longer accurate insofar as both the number of Thai inhabitants in Svalbard and the political situation have significantly changed in the past two decades.

services. As a result of the Svalbard Environmental Protection Act (Norwegian Ministry of Climate and Environment 2001), which protects all objects, structures, and sites from before 1946 as cultural heritage, Svalbard's history is also preserved as a visible material memory all over the archipelago. The mountain slopes around Longyearbyen are lined with the vestiges of former coal mines. In addition to the remains of coal mines, buildings, and other materials, the most striking example is the ghost town of Pyramiden in the Billefjord, accessible from Longyearbyen only by boat or snowmobile. This was one of the coal mining villages, founded by Swedish entrepreneurs in 1911 and sold to the Soviets in 1926 due to economic problems (Arlov [1989] 1994: 56). The Russians closed and abandoned Pyramiden in 1998. Preserved by the Arctic cold, the village sits on the waterfront just as it was when its human inhabitants abandoned it and is now populated only by birds, seals, and the occasional polar bear. Overlooking the Nordenskiöldbreen, one of Svalbard's largest but inexorably decreasing glaciers on the other side of the Billefjord, this dystopian scene now serves as one of Svalbard's most popular tourist attractions.

There is a certain irony to the fact that ghost towns bearing witness to bygone prosperous coal mining days, the few whales that have survived the extinctionist European whale hunt in Arctic waters, as well as melting glaciers and other slowly disappearing lifeforms, have now become tourist attractions – thus serving yet another extractive and environmentally harmful industry on Svalbard. This irony is less a paradox, though, than an illustration of a continuity between the various forms of human activity on Svalbard that have shaped and made the history of the archipelago. They all perform a resourcist, appropriative, and exploitative relation to the natural world that has made the 'natural environment' of Svalbard what it is today. The global depletion of natural resources and the steadily intensified burning of fossil fuels throughout the last couple of centuries, which also shape the making of more-than-human history on Svalbard through whaling and coal mining, are deeply implicated in the making of global existential threats to human and more-than-human life such as biodiversity loss and global warming (as well as other associated ecological problems).

The Arctic is one of the regions on earth where the already unfolding effects and challenges that come with these transformations and threats today are most visible. This has been proven throughout the last few decades by the growing research sector on Svalbard. Rising average temperatures cause glaciers to melt, permafrost to decrease, and rainfall to increase at the expense of snowfall, all of which affects the ecosystemic conditions of human and more-than-human life on Svalbard and the survival rates of non-human species (see e.g. AMAP 2012; Norwegian Polar Institute n.d.; Øseth 2011).

At the same time, with the establishment of the Global Seed Vault in 2008, Svalbard has now also become an iconic scene of the global struggle to counteract the loss of biodiversity. However, as I show throughout the following, this does not represent a break within the history of human activity on Svalbard. There rather exists a continuity between the activities of the Seed Vault and the larger system of agrobiodiversity conservation it represents and reinsures, on the one hand, and the extractive history of the archipelago on the other. This continuity lies in the appropriative resourcist relation to the natural world that has driven the depletion of the world's natural resources throughout history. This relation to the natural world is a contributing factor to the production of contemporary social-ecological crises including the large-scale loss of global agrobiodiversity *and also* shapes the endeavour of *ex situ* agrobiodiversity conservation performed and safeguarded by the Svalbard Global Seed Vault, if in a different form. I unfold this argument by analysing the history of Svalbard as a history of different forms of frontier making.

A History of Frontierism

Alex[7] is one of the first and few locals I meet in Svalbard. One day, they take me on a one- or two-hour tour around Longyearbyen and its surroundings in their truck, showing and telling me all kinds of things they know about Svalbard.

The landscape we pass through is marked by remains: snow that the summer temperatures haven't melted yet; cadavers from reindeer that haven't made it through the winter, now gnawed by the occasional Arctic fox; coal dust that the wind has blown around the island from the big pile sitting at the port; no longer active coal mines whose vestiges disrupt the scenic view of the mountainscape around Longyearbyen.

I ask Alex what the community is like, what kind of people live in Svalbard.

"People are selfish, but at the same time very friendly. They come here because they want to, not because they have to; if they don't want to be here anymore, they just leave. Everything is about making money. The biggest industry here, next to coal and research, is tourism; and frankly, people milk tourists like cows. But to be fair, tourists do come here

7 To protect the anonymity of the informant paraphrased in this vignette, I have replaced their proper name with a gender-neutral pseudonym and use the gender-neutral pronouns they/them throughout the text.

to have an exotic experience and are willing to pay a lot of money for it, so... The system is profitable, so people have no interest in changing it.

Almost no one is really from here and no one really stays very long. People come to live here for a few years, then they leave. Svalbard is like a no man's land; it is governed by Norwegian law, but at the same time, it's not Norway, everybody can come and live here; I mean, everyone from the countries that have signed the Svalbard Treaty but that's pretty much every country in the world. In a way, Svalbard is like a model for life on Mars; the Svalbard Treaty, it's like a model for treaties for new places.

I guess people come here because it's exotic. Many have been everywhere, and this is the last place for them to explore, like the last frontier. It's maybe similar to what the Wild West used to be in the US, you know, for the Gold Rush."

Alex's description of Svalbard and its population employs a remarkable array of metaphors and comparisons that mobilise capitalist and colonial narratives of world-making and resonate in many ways with the history of Svalbard as I have delineated it along the storyline of the Svalbard Museum. Reading the history of Svalbard through these metaphors enables me to develop an analytical perspective that foregrounds the relations to the natural world underlying the making of Svalbard throughout its history. This, in turn, makes it possible to situate the Svalbard Global Seed Vault within the more-than-nonhuman environment that Svalbard is by scrutinising continuities and discontinuities in the relations to the natural world it enacts. Two concepts that structure Alex's narrative are central to the following analysis: the imaginary of the Arctic archipelago as *terra nullius* that Alex invokes by describing Svalbard as a "no man's land", and their characterisation of Svalbard as "the last frontier".

First, the description of Svalbard as a "no man's land" is noteworthy considering that the archipelago's legal status changed a whole century ago. However, Alex's narrative is not primarily of interest in terms of its legal adequacy but in terms of the underlying ontological conception of land and nature and the associated relation to land and nature, which legal adjustments do not necessarily affect. There are countless philosophical and social-scientific accounts of the *terra nullius* imaginary that might ground such a discussion. The anticolonial and Indigenous studies scholars Tiffany L. King (2019: 68–69) and Max Liboiron (2021: 38–80) critically remark that much of this scholarship overemphasises land, or rather an understanding of land that fails to recognise a crucial prerequisite for casting land as *terra nullius*, namely its dehistoricisation. Dehistoricising land desired for appropriation means erasing its history, which usually means the displacement of Indigenous peoples and often includes dehumanising or genocidal violence against Indigenous and other racialised groups, as history has proven not least in the 'Wild West' Alex alludes to.

This intervention, which highlights the intersection of more-than-human with intra-human forms of violent appropriation, exploitation, and extermination, is a contribution to the scholarly discourse on the idea of land as *terra nullius* and its social implications, the importance of which cannot be overstated. With regard to Svalbard, it raises an interesting question. Since there are no historical records of Indigenous human life on Svalbard prior to the arrival of the Europeans (see footnote 2 in this section), the colonisation and exploitation of the archipelago and its resources are not built on a history of violent dispossession, subjugation, displacement, or genocide of native human societies. For some, therefore, it may stand to reason to conceive Svalbard as an actual *terra nullius* and its appropriation as a much more innocent process. I certainly do not want to suggest that there are no essential differences between intra-human and more-than-human forms of violence, exploitation, and subjugation. Notwithstanding that, in what follows I take up the ideas of ecofeminist philosopher Val Plumwood (1993) in analysing the ontologisation of land as *terra nullius* as a specific type of relation to the natural/ised world that is structurally analogous to other forms of othering.

In her momentous contribution to feminist theory and the philosophy of nature, *Feminism and the Mastery of Nature*, Plumwood (1993) argues that the notion of *terra nullius* constructs land as "a resource empty of its own purposes or meanings, and hence available to be annexed for the purposes of those supposedly identified with reason or intellect, and to be conceived and moulded in relation to these purposes" (1993: 4). On the basis of a critical engagement with the European philosophical canon, she argues that the *terra nullius* imaginary has historically established an ontological principle justifying the colonisation, appropriation, and instrumentalisation of land and nonhuman as well as human life. The foundation of this principle is a logic of dualism that assimilates the other/ed into the self by transforming it or them into "a mere tool or means for ends that are not its own" (1993: 192). By proposing and developing the foundations for an integrated analysis of human and more-than-human domination, Plumwood was one of the pioneers of a radically relational thinking about human-nature relations that informs much current scholarship on the matter, including my own approach. In what follows, I argue that it is the appropriative and exploitative relation to what is othered and desired as natural resource that has historically made Svalbard a *terra nullius* and a frontier of colonial and capitalist world-making. Insofar as this mode of world-making has significantly contributed to the emergence of the ecological crises of the present, it is crucial to consider the continuities between such destructive modes of world-making and the technologies and practices that promise solutions for their effects – as *ex situ* conservation does with regard to agrobiodiversity loss.

Second, Alex's characterisation of Svalbard as "the last frontier" resonates in a number of ways with the Svalbard Museum's narration of the archipelago's history in terms of the various extractive industries that have depleted Svalbard's natural resources – from seventeenth- and eighteenth-century whaling, through the hunting and trapping of other animals, to the nineteenth- and twentieth-century coal rush. Comparing Svalbard to "what the Wild West used to be in the US", Alex invokes a classical notion of the frontier such as articulated in "the proud frontier story of the making of 'America'" (Tsing 2005: 58) through the conquest and civilisation of spaces constructed as priorly wild and empty. In this story, the frontier is a spatial category denoting the moving border between 'wild' and 'civilised' spaces and a social category marking the assumed qualitative difference between them. Insofar as it obfuscates the violent dispossession and displacement of Indigenous peoples and the despoilment of environments underlying the history of nation-making through conquest and colonisation, frontierism is a partial – and certainly not impartial – story told from the perspective of the conqueror. Consequently, the notion of the frontier can be (and has been) challenged by foregrounding the perspective of the dispossessed and displaced. This is not possible in the case of Svalbard. Scholars in critical political economy and political ecology have developed a different kind of critique that is more suitable for a critical discussion of Svalbard's history of frontierism, which scrutinises and differentiates the notion of the frontier as an analytical category. This body of scholarship makes it possible to expose the ties between the Svalbard Global Seed Vault, with its mission of *ex situ* conservationism, and this history.

The environmental historian Jason Moore "operationalize[s] the concept of the frontier for the study of world capitalist expansion and its structural tendency towards environmental degradation" (2000: 409). While Moore builds on world-system theory, he also criticises its narrow understanding of the frontier "as a zone of incorporation" (Moore 2000: 410; see also 2015). According to Moore, in a world-system analysis of capitalist expansion as a historically and spatially expanding process of capital accumulation driven by the exploitation of ever more labour power, land, and natural resources, the notion of the frontier denotes the moving border between already and not-yet capitalised spaces, systems, and modes of production as well as the process of capitalist incorporation that moves this border (see e.g. Hopkins et al. 1987; Wallerstein 1989). Using the example of sugar cane, Moore argues that capitalist expansion is driven, at its heart, by "the production and distribution of *specific* commodities" (Moore 2000: 410, emph. in orig.) and the extraction of the raw materials they are made from. To encapsulate this specification, he introduces the notion of the *commodity frontier*. Insofar as commodity frontiers are always and inherently place-specific, the concept entails an understanding of frontierism as a "mode of capitalist expansion" (Moore 2000: 411, emph. in orig.) – here through the transformation of raw materials into commodities. According to Moore, then, the frontier is not "a certain kind of place" so much as "a certain kind of socio-spatial movement" (Moore 2000: 412). As such, commodity frontiers are not a result of the consolidation of global capitalism so much as its precondition and modus operandi (Moore 2000: 428).

Analogously, the environmental degradation that comes with the depletion of raw materials for ever-expanding commodification is not an unintended consequence so much as the source of new commodity frontiers, and hence a structural component of the frontier mode of capitalist expansion:

> "The extractive and agricultural character of frontier industrialization under conditions of ceaseless capital accumulation meant that not only was ecological exhaustion a fact of life in these areas, but that ecological exhaustion was a major impetus to further capitalist expansion and to the system's cyclical fluctuations." (Moore 2000: 412)

This is pertinently exemplified by the loss of plant genetic diversity in the wake of the Green Revolution and the concomitant expanding interest of industrial agriculture and the seed industry in regions with high biodiversity, often land cultivated by peasant farmers, as well as by the creation of a market for agrochemicals. In Svalbard, the history of whaling and later extractive industries illustrates Moore's argument. It was the exhaustion of the whale populations in Labrador and Newfoundland that first drove whalers to Svalbard's inshore waters and, after their exhaustion, to the open waters around the archipelago. With the near extinction of Svalbard's whale population other commodity frontiers emerged, which meant that capital accumulation in Svalbard was sustained by the exploitation and exhaustion of other resources. It is in this dynamic, Moore argues, that "[e]cological exhaustion at the point of production [is] complemented by an environmentally destructive multiplier effect" (2000: 412).

Moore's discussion of commodity frontierism makes an important contribution to world-system theory's discussion of frontierism by offering a comprehensive and place-specific analytical tool for understanding historical and ongoing capitalist expansion through commodification and its correlation with environmental degradation. He seeks to foreground "the link between ecological transformation and the expansionary logic of world capitalism [...] to suggest ways of rethinking early modern capitalist expansion as a socio-*ecological* process" (Moore 2000: 409, emph. in orig.). In *Capitalism in the Web of Life: Ecology and the Accumulation of Capital* (2015), Moore further elaborates his attempt to counter "Capitalism's governing conceit [...] that it may do with Nature as it pleases, that Nature is external and may be coded, quantified, and rationalized to serve economic growth, social development, or some other higher good" (2015: 14) with a non-dualist analysis of the social-ecological historical reality of what he calls the "web of life". While this project conclusively historicises the capitalist valorisation of the dualist conception of society and nature, it remains overdetermined by an understanding of natural-cultural history as mainly driven by political economy. Thus, for one thing, Moore overlooks the importance of political relations and processes (see also Battistoni 2017: 29 fn 58). For another, although he highlights the links between and interdependence of society and nature, he defaults to an anthropocentric perspective on social-ecological processes that does not question a resourcist paradigm. This paradigm, however, lies at the heart of the *terra nullius* imaginary and the appropriative frontierism(s) it paves the way for.[8]

Anna L. Tsing (2005)[9] articulates a similar critique, based on which she develops a notion of the frontier rooted in the frameworks of multispecies

---

8 Supporting Moore's effort to articulate analytical categories that include both human and non-human productivity in economic processes, yet critical of his theorem of "cheap nature" (Moore 2015), Battistoni develops a compelling conception of "the 'work of nature' as a collective, distributed effort of humans and nonhumans acting to reproduce, regenerate, and renew a common world" (2017: 5). I thank Vicky Kluzik for pointing me to the work of Battistoni and Andreas Folkers for thought-provoking discussions of Moore's work.

9 The text cited here was first published as "Natural Resources and Capitalist Frontiers" in *Economic and Political Weekly* two years earlier (Tsing 2003). The 2005 version I refer to here is slightly updated but largely the same.

ethnography and political ecology. The aim of her rearticulation of the frontier is to abandon the premises and problematic implications of a resourcist paradigm: "Most descriptions of resource frontiers take the existence of resources for granted. [...] The landscape itself appears inert: ready to be dismembered and packaged for export. In contrast, the challenge I have set myself is to make the landscape a lively actor." (Tsing 2005: 53) Instead of starting from the premise that nature is a given set of resources, Tsing asks how nature becomes a set of resources on or through the frontier. This question comes with an understanding of the frontier as "not [...] a place or even a process but an imaginative project capable of molding both places and processes" (2005: 59). Frontiers, then, are "both material and imaginative" (Tsing 2005: 53).

While Tsing does not elaborate on her conception of the imaginative, her arguments resonate with how the anthropologists Sophie Chao and Dion Enari (2021) discuss dominant and transformative climate change imaginaries. Following their line of argument, in what follows, I understand the imaginative as referring to "the varied and complex ways in which realities are conceptualised, fashioned, and contested [...] [which are] shaped by divergent socio-cultural contexts, material-ecological terrains, and ideological and physical infrastructures that link the local to the global through partial and uneven connections" (Chao/Enari 2021: 35). Against this background, conceiving the frontier as an imaginative project indicates that it matters (to put it in a Harawayian manner), both in the sense that it makes a difference and that it has very real material consequences, what human and more-than-human worlds imaginations of current and future worlds are situated in (see also Tsing 2015). Put differently: the frontier as an imaginative project is a material-semiotic world-making device.

Rather than merely historicising imaginations of the social and the ecological, whether as separate (as in capitalism) or interdependent spheres (as in Moore's web of life), Tsing's ethnographical analysis traces the entangled material and imaginative practices that make social-ecological worlds. For her, it is in such entangled material-semiotic becoming that frontiers come to life in locally specific and therefore multiple forms (2005: 68). Tsing, thus, differs from Moore by seeing neither the time-spaces of transformation (e.g. zones of incorporation) nor the transformations themselves (e.g. practices or processes of incorporation) as what constitute a frontier. Rather, for her, the frontier emerges in the nexus of desire and promise evoked by and at the same time "encouraging ever-intensifying forms of resourcefulness" (Tsing 2005: 59). Two questions that Tsing raises in this context are central to my inquiry into the imaginary of Svalbard as "the last frontier" and the question of what role the Seed Vault plays in this imaginary: first, *what forms* the frontier assumes in the constantly evolving terrain in which it emerges, and second, *how* (and to what extent) the desire-driven project of frontierism turns nature into a pool of resources (see Tsing 2005: 54–60). I proceed by unfolding the ways in which the history of Svalbard is a history of multiple frontiers making Svalbard what it is. Tracing these different frontiers, then, allows me to address the second question by asking what the different forms the frontier assumes in Svalbard have in common.

Probably the most commonly known form of the frontier, and the first form in which it came to Svalbard, is what Tsing refers to as the *nation-making frontier* (2005: 57). This goes back to the history of European colonialism and to colonial stories of the making of the Americas through a progressive process of 'civilization' of formerly 'wild' spaces (Tsing 2005: 57–58, quoting Sarmiento [1845] 1998 and Turner [1893] 1994). When the nation-making processes of spatial expansion and resource exploitation in the Americas and beyond were stretched to their limits (for the time being), Tsing argues, "frontier optimism" as the main sentiment associated with frontierism was replaced by "frontier nostalgia" (2005: 58). In the aftermath of World War II, however, frontier optimism experienced a renaissance in light of the promises of industrial technology. The start of a new era of seemingly unlimited resource extraction and capitalist expansion gave rise to a new type of frontier Tsing calls the *technofrontier* (ibid.). The new technofrontier mode of capitalist expansion, as it emerged in the mid-twentieth century, operates according to a new system of coordinates not only spanning the entire globe but also penetrating ever deeper into the earth and its resources. At the same time, the technofrontier preserves the imaginary of 'civilising' formerly 'wild' spaces in developmentalist narratives, thus maintaining relations with "its older frontier cousins" (Tsing 2005: 58).

The history of Svalbard as I have delineated it above is a condensed and illustrative example of frontier-making. Svalbard first became a frontier in the European colonial project when Dutch seafarers exploring Arctic waters in search of a Northeast passage to Asian markets first landed on the archipelago in the sixteenth century. In the subsequent struggles over its territory and natural resources, Svalbard became part of European and Russian (and then Soviet) histories of nation- and empire-making. The commodity frontier, if one were to think with Moore, shaped various extractive industries exploiting the archipelago's resources beginning with whaling. With the Arctic coal rush of the late nineteenth and early twentieth century, the technofrontier came to Svalbard. Unlike in the story of a period of frontier nostalgia before the return of the frontier in the form of the technofrontier, however, in Svalbard nation-making and technofrontierism were entangled endeavours at least until the adoption of the Svalbard Treaty in 1925 resolved the legal status of the archipelago. Until the coal mining industry was plunged into crisis and nationalised in the 1970s, after which it went into continuous decline, the technofrontier mode of capitalist expansion through technologically driven exploitation of natural resources remained the primary force shaping human activity on Svalbard.[10] In the second half of the twentieth century, the promises of limitless capitalist expansion through the technological penetration of the earth began to unravel, and not only in Svalbard.

At that time, Tsing (2005: 58) argues, it became increasingly undeniable around the world that the global advancement of the technofrontier was contributing significantly to accelerating ecological devastation. The ecological ravages of technofrontierism marked the birth of an international environmentalism that soon took the shape of measures for global environmental protection and conservation. Again, the history of the Green Revolution, its social-ecological repercussions, and the conservationism emerging in response to them exemplifies this (see chapter 1). According to Tsing, however, the environmentalism of the late twentieth century did not so much challenge frontierism as translate the technofrontier into yet another new form: "This is the salvage frontier, where making, saving, and destroying resources are utterly mixed up, where zones of conservation, production, and resource sacrifice overlap almost fully, and canonical time frames of nature's study, use, and preservation are reversed, conflated, and confused." (Tsing 2005: 59)

Tsing develops the notion of the *salvage frontier* through an ethnographic study of logging practices in Kalimantan, Indonesia, which highlights some of the many ways in which processes of saving and destroying environments are inextricably entangled.[11] She argues that frontierist dreams of endless capital-

10 The contemporary resurgence of imperial conflicts raises the question of whether the nation- or empire-making frontier might experience a comeback on Svalbard (see e.g. Waehler 2022). While this is not the place for a discussion of this question, it is worth noting that this development indicates that the history of the frontier is less sequential than genealogical and that different forms of frontier are not mutually exclusive.

11 Tsing deals with the notion of the salvage frontier fairly briefly in this text and does not revisit it elsewhere either. To my knowledge, it has not been picked up or systematically developed further by anyone else. Accordingly, my summary of the concept is also fairly short. Annukka Paajanen's (2024) analysis of the Svalbard Global Seed Vault makes a

ist expansion come with a vested interest in measures of environmental protection and their promotion insofar as they harbour "fears of apocalypse" (Tsing 2005: 68). However, environmental protection and conservation measures usually only cover a small portion of the land or resources threatened with destruction or extinction, whereas the ecologically devastating appropriation of land and resources that has generated these threats in the first place continues elsewhere. In the case of logging and other large-scale land use practices that destroy whole environments, a common strategy of environmental protection is the establishment of reserves: "The reserve model [...] arises in places where environmentalists are panicked by the possibility of total destruction; it argues that *something* must be saved" (Tsing 2005: 69, emph. in orig.). As Tsing illustrates through the example of the Kalimantan frontier, in many cases this type of selective environmental protection ultimately works to save the frontier mode of capitalist expansion itself insofar as resource companies trade off (small) protected areas against rights to deplete other (large) areas. If there is a silver lining, she argues, it is "the mobilization that it [the reserve model; FV] has inspired, which brings the possibility of citizenship to those who never had them [sic] before: small collectors, tribes, trees" (Tsing 2005: 69). In other words, there is a possibility that established orders of othering and exploiting the natural/ised world may be destabilised in the wake of environmentalist mobilisation.

Although Tsing's conceptualisation of the salvage frontier is anchored in the examples of logging in Kalimantan and environmental protection through reserves, the concept is not determined by this context. As elaborated above, Tsing conceives the frontier as "a globally travelling" (2005: 68) imaginative project that comes to life by shaping places and processes in local and therefore multiple contexts and forms. Applying this idea to the Svalbard Global Seed Vault and the *ex situ* approach to conserving endangered agrobiodiversity illustrates that it is not only reserves and other *spaces* of *in situ* conservation that the salvage frontier moulds, but also *forms and processes* of conservation. As I have shown in the first chapter and continue to discuss throughout those that follow, the alarmed awakening to the disastrous consequences of irreversible loss of plant genetic diversity in the second half of the twentieth century prompted a rush to save "something" by conserving seeds in genebanks. In the

similar reference to Tsing (2015) to conceptualise the Seed Vault as "a site of salvage accumulation" (Paajanen 2024: 23–25). However, the notion of the salvage frontier, which is of particular interest to me here, is not addressed in her work.

course of this attempt, *ex situ* conservation – the key international strategy to preserve plant genetic diversity for food and agriculture – moulds what agrobiodiversity is currently and will be in the future. By organising conservation around "plant genetic resources" and by selecting which plant material is worth conserving, whatever the rationalities of selection, *ex situ* conservation consolidates a geneticised and technologised as well as a resource-oriented understanding of agrobiodiversity – the technofrontier version of agrobiodiversity.

Importantly, the preservation of this kind of agricultural biodiversity, especially in genebanks with a "productivist" or "preservationist profile" (Pellegrini/Balatti 2016: 2762–2764; see chapter 1.5) such as the Seed Vault and its depositors, is not an end in itself. It serves to make available the largest possible stock of plant genetic resources not only for future farming, but also for the breeding of new varieties supposed to help adapt the existing system of global industrial agriculture to future social-ecological challenges. In this sense, *ex situ* conservation of PGRFA and the Svalbard Global Seed Vault as the stronghold of this practice and understanding of agrobiodiversity operate in the salvage frontier mode of capitalist expansion – which ultimately also preserves the technofrontier mode. The crux of this is that rather than challenging an agricultural system that is built on technofrontierism and therefore complicit in the emergence of the social-ecological challenges it faces, *ex situ* conservation serves to sustain that same system of global industrial agriculture. By doing so, however, it also sustains the roots of the agrobiodiversity crisis it is supposed to be a response to.

As an iconic manifestation of the salvage frontier, then, the Seed Vault is the most recent iteration of Svalbard's history of frontierism. This history is not a series of consecutive phases throughout which the frontier assumes alternating, mutually exclusive forms; it is not, as one might conclude, "a series of ironic twists" (Tsing 2005: 59, quoting Schmink/Wood 1992). Instead, Tsing emphasises that "[s]uch twists are more than irony: they predict and perform their own reversals, forming productive confusions and becoming models for other frontiers" (Tsing 2005: 59). As I have shown through the history of Svalbard, different forms of the frontier can coexist and permeate one another. This genealogical understanding of the history of frontierism, illustratively condensed in the history of Svalbard, lies at the heart of my effort to situate the Global Seed Vault in Svalbard as its home or more-than-nonhuman environment. The history of Svalbard epitomises the emergence of the salvage frontier out of the dreams and desires associated with the technofrontier and in response to the destructive forces it has unleashed. In like manner, the Svalbard Global Seed Vault was established to counteract the devastations and threats that have come to matter through the Green Revolution and industrial agricultural production more generally, while also preserving the technofrontierist dreams and promises associated with them.

Before going on to elaborate in more detail on the practices and politics of *ex situ* conservation and the Seed Vault's role in them, I want to conclude this chapter by reflecting on the second question raised earlier: How does nature become a set of resources in and through the frontier(s)? I address this question by tracing the making of resources through the different forms of frontierism and looking at what they have in common.

Thinking the frontier through Tsing as an imaginative-material project, or in Harawayian terms as a material-semiotic mode of world-making, means understanding frontierism not just as a mode of appropriating land and natural/ised resources but as a particular way of being in and knowing the world that *makes* land, nature, and biological diversity a set of resources. To develop this claim, I will briefly turn to a region that is geographically distant but closely connected to my argument through the history of frontierism. The frontier, in most of its various forms described above, creates what Macarena Gómez-Barris (2017) calls "extractive zones". Underlying her conceptualisation of this notion is a study of different regions in South and Central America that have a long and intertwined history of exploitative resource extraction and repression of local Indigenous communities. These areas, Gómez-Barris argues, become extractive zones in the encounter with "the colonial paradigm, worldview, and technologies that mark out regions of 'high biodiversity' in order to reduce life to capitalist resource conversion" (2017: xv).

The process of converting areas rich in a diversity of natural resources into extractive zones for colonial capitalism commenced with the European colonial expansion that began in the 1500s. Londa Schiebinger's (2004, 2005) historical research on colonial botany shows that a key motive for many of the explorative voyages of the colonial era, beyond territorial expansion, was prospecting for and extracting natural resources considered valuable for European markets. While Svalbard became a temporary and globally rather insignificant extractive zone for oil extraction from whale and seal blubber and later for coal mining, South and Latin America are home to a number of extractive zones historically much more significant on a global scale. This is due to the fact that a number of the centres of plant and crop diversity (Vavilov [1926] 1992a; Khoury et al. 2016) are located in these parts of the world. They became extractive zones through the colonial appropriation of the "green gold" (Schiebinger 2004: 7) that flourished there; more precisely, through the appropriation of botanical resources as well as Indigenous knowledge about their cultivation and the transplantation of these resources from the 'New World' to botanical gardens and plantations in the 'Old World' and its colonial satellites (Brockway 2002; Philip 2004; Schiebinger 2004; Schiebinger/Swan 2005). Alfred W. Crosby famously encapsulated the global movement of plants and other biological resources that began in the colonial era in the notion of "the Columbian exchange" (1972). Countering the euphemistic tone of this expression, others more critically refer to this process and the new forms it has been assuming throughout history as "biopiracy" (Shiva 1998; Kloppenburg 2000; Nehring 2022) or "biocolonialism" (Whitt 1998; Hawthorne 2007). The historical processes these authors and terms describe illustrate how nation- and empire-making, as well as resource and commodity frontiers, intertwine in the creation of extractive zones and the transformation of biological diversity into a resource for the thriving of empires.

A crucial argument in both Schiebinger and Gómez-Barris is that the practices of extractive colonialism co-emerged and are deeply entangled with the epistemological pillars of modernity. Schiebinger's historical analyses foreground the ways in which the emergence and propagation of modern science and its ideal of objective and observer-independent knowledge are interwoven with the European colonial powers' seizure of economic and geopolitical power between the fifteenth and the nineteenth century (see also Philip 2004; Schiebinger/Swan 2005).[12] In line with this, Gómez-Barris argues that the epistemological principles of the dominant modern sciences are both precondition and product of the extractive exploitation and commodification of land, nature, and peoples conceived as resources available to be harnessed for modernist 'civilisational' purposes:

> "Before the colonial project could prosper, it had to render territories and peoples extractible, and it did so through a matrix of symbolic, physical, and representational violence. Therefore, the extractive view sees territories as commodities, rendering land as for the taking [...]. This viewpoint, similar to the colonial gaze, facilitates the reorganization of territories, populations, and plant and animal life into extractible data and natural resources for material and immaterial accumulation." (Gómez-Barris 2017: 5)

12 I unfold this argument in more detail in chapter 4.2.

The construction and utilisation of biological material as a set of natural resources thus becomes discernible as an *onto-epistemological* project: epistemology and ontology, knowing and being, are inseparably entangled (see also Barad 2007: 379, 409). Insofar as the extractive *view* – much like "the resourcist cosmovision" (Fenzi/Bonneuil 2016: 78; see chapter 1) – informs extractivist *practices* and *modes of being in the world*, it is ontologically implicated in the world that comes to matter in these onto-epistemological practices. To put it in Harawayian terms: It matters what stories, imaginaries, and knowledges make worlds and what worlds these, in turn, emerge from.

Thus understood, a problem with both the notion of the extractive view and the resourcist cosmovision crystallises. They each denote a particular *perspective on* the world that guides *interactions with* the world, thus implicitly positing the world as an entity at least to some degree ontologically preexisting its perception and moulding. As I have shown throughout this chapter, however, resourcism, extractivism, and frontierism have shaped the becoming of the world in a very real, material sense that is not reversible. They are modes of "material-semiotic worlding" (Haraway 2016: 13): Where frontiers come to life in the nexus of dreams and desires of endless capital accumulation, extractive zones emerge in and continuously reproduce the material-semiotic nexus of othering and appropriation that allows frontier dreams to become realities. Land *becomes terra nullius* and biological materials *become* natural resources through their construction and production as part of an inherently purposeless sphere that is ontologically different from the human, as resources available to be appropriated and exploited for the purposes of the appropriator.[13]

In Svalbard, the various resource frontiers that turned the archipelago into an extractive zone to benefit the making of European nations and empires have driven the whale and other animal populations to the brink of extinction. The technofrontier that came to Svalbard in the form of the coal mining industry

13 Scholars emphasising the structural analogies between human and more-than-human forms of othering and instrumentalisation argue that the same logic renders certain human lives exploitable, enslavable, and killable along racialised, gendered, classed, and other socially stratified lines of othering by way of excluding them from the sphere of the properly human. These analyses understand the human as a historical rather than an ontological category, invented by those "who defined themselves as humans in their praxis of living and applied their self-definition to distinguish and classify and rank lesser humans" (Mignolo 2018: 153; see also Plumwood 1993; Nath 2009; Jackson 2020).

has depleted the fossilised biodiversity inside Svalbard's mountainscape, fuelling not only the making of economic empires but also global warming, which has become one of the greatest contemporary challenges to ecological stability not only in Svalbard. In Svalbard as elsewhere, the frontier mode of worldmaking has created a world of disappearing more-than-nonhuman ecological and more-than-human social worlds. With growing public awareness of the social-ecological threats arising from global climate change and other (related) ecological hazards, such as the increasing vulnerability of plants and ecosystems and the advancing loss of biodiversity, conservationist efforts have created a new type of frontier. The salvage frontier comes to matter on Svalbard in the form of a "cold rush" (Pietikäinen 2024) in search of a new kind of resource: the conservational capacity of the Arctic cold and permafrost and the promise that they will be able to conserve the green gold of the past for the future. Salvaged in the Arctic permafrost inside the Svalbard Global Seed Vault, then, are the "plant genetic resources" the technofrontier mode of modern agro-industrial expansion has created along with the frontierist dreams of sustained capitalist expansion they carry.

3.2 The Svalbard Global Seed Vault (II): A Vault of Many Worlds

> Hundreds, maybe even thousands of journalists and other people have written articles about Svalbard [the Global Seed Vault; FV] and virtually none of them has gotten it right (laughs). And I realised [...] that I thought the story as I knew it needed to be written down. (Cary Fowler, 19/20 March 2021: l. 88–96)

Probably the most popular, widespread, and at the same time fallacious stories told about the Svalbard Global Seed Vault are those that depict it as a "doomsday vault" or a "Noah's Ark" for the world's seeds. Coined by international media reporting on the Seed Vault after it first opened, these metaphors quickly gained momentum and subsequently informed other stories, including scholarly accounts of the Seed Vault (Qvenild 2008; Fava 2013; Gan 2015; Nadim 2018) as well as the public relations work of the partners behind the Seed Vault.[14]

14 In the early years after its opening, the partners behind the Seed Vault played their part in reproducing the doomsday narrative, possibly to attract international attention and thus further funding and seed deposits. The Crop Trust, for instance, tagged early

However, as Cary Fowler, the founder of the Seed Vault, frequently emphasises, these popular narratives "typically mask the complexity of the endeavor and, if anything, underestimate its practical utility" (Fowler 2008b: 190). While the Seed Vault is designed to be able to conserve seeds "for hundreds of years, or longer" (Fowler 2016: 18), contrary to the story of the doomsday vault, this long-term orientation does not derive from the anticipation of a global apocalyptic event in the wake of which agriculture and civilisation can be restored with the help of the Seed Vault. Instead, the aim of conserving the greatest possible crop diversity in Svalbard for as long as possible is to keep these resources available for global agriculture to adapt to changes and challenges in both the near and the more distant future (Fowler 2016: 17–18).

Interestingly, in his efforts to counter the doomsday story by underlining the very mundane and practical use of the Seed Vault, Fowler often resorts to a remarkably messianic rhetoric characterised by a strong salvationist pathos.[15] In a short piece about the Seed Vault published the same year it was opened, Fowler stated that "[c]onserving crop diversity is the prerequisite for the future evolution and success of agriculture. Few other human problems will be solved, if this one is not" (2008a: 501). In *Seeds on Ice: Svalbard and the Global Seed Vault* (Fowler 2016), the book Fowler wrote to counter the often inaccurate media coverage with a comprehensive account of the Seed Vault, he describes it as a "remarkable effort to save the past and the future of agriculture" (2016: 18; see also Fowler 2008c).[16] This effort is the result of Fowler's life-long dedication to the conservation of endangered agrobiodiversity, around which he built his career.

After a period of activism in the civil society sector, Fowler held various academic positions before he eventually worked in a number of the major international agro-political institutions and arenas, where he played a key role in the process of establishing the Svalbard Global Seed Vault. Over the course of his

pictures of Seed Vault posted on *flickr* (an online community for sharing amateur and professional high-resolution photos and videos) with "doomsday vault" as a caption. I thank Marleen Boschen for sharing this observation with me.

15 In doing so, Fowler masks the ways in which the doomsday story does, in fact, matter to the Seed Vault. I discuss this, as well as the tension between the ways in which the doomsday narrative is both rejected and embraced by the people behind the Seed Vault, in chapter 5.1.

16 For a critical analysis of the media coverage of the Svalbard Global Seed Vault, see Chacko (2024).

career, Fowler has (co-)authored three books, each of which reflects the position from which it was written. The activist tone of *Shattering: Food, Politics, and the Loss of Genetic Diversity* (Fowler/Mooney 1990), co-authored with his former partner in civil society Pat Roy Mooney, later gave way to a more academic discussion of agrobiodiversity loss and conservation in *Unnatural Selection: Technology, Politics, and Plant Evolution* (Fowler 1994). *Seeds on Ice*, in contrast, which Fowler himself describes as "a love letter" (2016: 20) to Svalbard and the Global Seed Vault and which vividly interweaves the story of global crop diversity loss and the Seed Vault's conservation efforts with the story of Svalbard perishing in the light of global warming, reads like an ode to a world on the verge of extinction.

All three books open with a reference to the American geneticist and crop historian Jack Harlan, who first inspired Fowler to dedicate his career to saving crop diversity. The words he "will never forget" (Fowler 2009: 0:26), as Fowler put it in a TED talk a year after the opening of the Seed Vault, stem from an article with the expressive title "Genetics of Disaster" (Harlan 1972). In this paper, Harlan conceptualises the diversity within crops as a genetic resource that, as Fowler repeatedly quotes, "stand[s] between us and catastrophic starvation on a scale we cannot imagine. In a very real sense, the future of the human race rides on these materials" (Harlan 1972: 212, quoted in Fowler/Mooney 1990: 6, Fowler 1994: xiii, 2016: 5). For Fowler, against this background, the point of the Seed Vault is to collect and conserve "[v]irtually everything, every trait we might want our crops to have in the future – all the options – [that] can be found in this genetic diversity" (Fowler 2016: 17–18), so as to enable farmers and plant breeders to adapt their crops to increasingly uncertain future conditions (Fowler 2016: 14).

Continuing the exploration of the Svalbard Global Seed Vault as a central component of the international response to agrobiodiversity loss, this section discusses the Seed Vault's unique role in the global system of agrobiodiversity conservation as well as the rationality underlying it. To do so, I tell the story of the Seed Vault from the perspective of the partners behind it as well as other actors who have been involved in the past, such as the FAO and Bioversity International (formerly IPGRI, IBPGR). In addition, I discuss how critical commentators from the nongovernmental and civil society sector perceive the Seed Vault. Through passages from the expert interviews conducted for this study, I develop an account of what the Seed Vault was designed and envisioned to be, what it is and is not today, and what the interested public perceives it to be, to do, and to be able to be and do.

I begin by situating the creation of the Seed Vault in a larger historical and agro-political context, in order to explain its unique role within the world of agrobiodiversity conservation as backup for other genebanks. I then discuss the challenges the Seed Vault has faced when its own backup, the Arctic permafrost, which was supposed to provide the securest possible environment in the world for the long-term cold storage of seeds, turned into a threat to the seeds by melting and leaking into the facility. Sensationalist media reports presented this incident as proof that climate change had undermined the stronghold against disasters such as climate change. The expert interviews I conducted show that the story is more nuanced than that. The section concludes by discussing divergent interpretations of the political implications and impact of the Seed Vault. It is through these that I develop the analytical understanding of the Seed Vault I propose, namely as not only a *technology* promising to save the world by saving the seeds of more-than-human futures, but as a political *arena* assembling a diversity of worlds collectively negotiating the reality of agrobiodiversity loss and conservation.

Securing the Past and Future of Agriculture

As I showed in chapter 1, *ex situ* conservation became the globally predominant response to the increasing de-cultivation and concomitant large-scale loss of agrobiodiversity when the global agro-political community became aware of the pitfalls of this development conjured by the modernisation of agriculture. According to the Acting Secretary of the Seed Treaty, Kent Nnadozie, the primary focus on *ex situ* conservation as an international approach to counteracting the growing loss of plant genetic diversity gained acceptance in the second half of the twentieth century because it was considered to be the appropriate response to the perceived urgency of the unfolding agrobiodiversity crisis:

> "Because of the rapidly increasing rate of loss of diversity – in terms of habitat and land use changes and focus on monocropping – there was a lot more move and drive, a push to have lot more *ex situ* conservation, to collect and preserve what is there so that that's not lost in the process." (Kent Nnadozie, Secretary, Seed Treaty, 26 March 2020: l. 789–795)

As described here by Nnadozie, the historical emphasis on *ex situ* conservation was first and foremost an emergency measure taken due to the urgent need to find a response to the plant genetic erosion caused by the modernisation of global agriculture. While this logic is reasonable in terms of taking short-term security measures, it focuses on the symptoms rather than the roots of the problem, which the conservation of plant genetic diversity in genebanks leaves untouched. With the proliferation of *ex situ* conservation came the realisation within the international agro-political community that the conservation of plant genetic resources in genebanks, too, is not a failsafe measure against agrobiodiversity loss. Genebanks are constantly exposed to a variety of threats ranging from wars to storms and floods, as well as technical problems such as power cuts or malfunctions of refrigeration equipment (Fowler 2008c: 12–13).

This shows that agrobiodiversity loss is not an issue restricted to *in situ* environments but a problem that can equally occur *ex situ*. To give a few examples: In 1971, Nicaragua's national seed bank fell prey to an earthquake; in 1998, the national seed bank of Honduras was destroyed by a hurricane; the national genebank of the Philippines was flooded in the wake of a typhoon in 2006 and then again destroyed by a fire in 2012. Wars and uprisings have caused the devastation of seed banks in Liberia and Sierra Leone in the 1990s, and in Afghanistan and Iraq in 2001 and 2003. In most of these cases, the diversity conserved within these seed banks was irretrievably lost (Fu 2017).

Ex situ collections often include a range of varieties nowhere cultivated *in situ*; as a result, the loss of such collections can mean that these varieties go extinct altogether. Once this danger began to be appreciated in the 1970s, the FAO developed yet another security measure. Genebanks in the member states were urged to create safety duplicates of their seed collections, from which these could be rebuilt in case of loss (Curry 2022b: 665). At that time, it was already common for genebanks to have dormant backups of their active collections, usually stored either in the respective genebank itself or by a partner institution. To prevent irretrievable losses of agrobiodiversity due to catastrophic events such as mentioned above, the new safety duplication standard required genebanks to create a second copy of the collection that would serve as a safety duplicate and should therefore be stored in a different, preferably geographically distant location. While the more well-equipped seed banks soon amended their standard procedures, part of the work done by the Crop Trust and its partners to this day involves supporting smaller and less well-equipped genebanks in applying this standard.

The latest version of the FAO's *Genebank Standards* (FAO 2014) recommends that safety duplicates should be stored under the best possible long-term storage conditions and, if possible, in a different country. The primary factors of what makes a location optimal for storing safety duplicates are defined as follows:

> "Facilities must ensure low radiation (radioactivity) and stability (low probability of earthquakes). The facility must be situated at an elevation that guarantees proper drainage during seasonal rains and eliminates the risk of flooding in the event of rising sea levels due to global warming. Equally important is economic stability and socio-political certainty. Koo et al. (2004) suggest that safety duplicate samples should be located away from the risk of political embargo, military action or terrorism that could disrupt international access." (FAO 2014: 58)

The Svalbard Global Seed Vault was designed to facilitate safety duplication under exactly these conditions – to provide, as Fowler puts it, "the ultimate safety net for existing genebank collections" (2008a: 501). Accordingly, the latest edition of the *Genebank Standards* explicitly recommends the Seed Vault (and no other facility) as an internationally accessible facility offering the kind of storage space required for safety duplicates.

Herein lies a crucial characteristic of the Seed Vault that essentially distinguishes it from other genebanks. A scientist from the Crop Trust, Hannes Dempewolf, explains the unique role of the Seed Vault compared to other genebanks as follows:

> "It's a global backup for seed banks all around the world. There are a lot of misunderstandings around it. People think that is what a genebank is, which is totally wrong and something that we are constantly trying to fix when we talk to media about it; making people understand that there are seeds that are alive, that you have to regenerate them. This is not a solution where you just shove something and then dig it up 200 years later. This is something that is a living, breathing system." (Hannes Dempewolf, Scientist, Crop Trust, 28 May 2019: l. 1031–1041)

This interview passage shows that the conservation of crop plants in genebanks cannot be reduced to the cold storage of boxes filled with seeds. This type of conservation is restricted to facilities with a "preservationist profile" (Pellegrini/Balatti 2016: 2763–2764; see chapter 1.5), such as the Svalbard Global Seed Vault. The seeds it stewards, in contrast, come from "working collections" (Curry 2017b: 3, 2022b: 665) actively conserved in genebanks with a "productivist profile" (Pellegrini/Balatti 2016: 2762–2763). As chapter 4.1 elaborates in more detail, the work of conservation performed in genebanks of the latter kind involves a whole series of practices and technologies that make it possible for seeds to remain viable in long-term storage. The active and ongoing work of conservation includes drying and packaging seeds in air-tight bags for long-term storage as well as periodic revitalisation, regeneration, and renewed conservation. The exclusive and exceptional purpose of the Seed Vault, in contrast, is to securely store the packaged and boxed safety duplicates of genebank collections for the long term. Crucially though, although the Seed Vault has a different conservation profile and purpose than the genebanks whose safety duplicates it holds, the interview passage also highlights that they are interdependent parts of a larger system of crop conservation. While it is the genebanks around the world that do the breathing, the Seed Vault fulfils a crucial role in keeping the "living breathing system" of crop conservation alive.

The first, and so far the only, case in which the Seed Vault came into effect in the sense that a depositor had to withdraw their deposit in order to reestablish a lost seed collection is the case of ICARDA in Syria outlined in the introductory chapter. Through the example of this case, a scientist working with the FAO elaborates the role of the Seed Vault within the world of conservation as follows:

> "It is a safeguard and it has its importance in that it is there in the event of a disaster – which happened at ICARDA. And it's a success story; maybe not as big as the news makes it out, but it's a success story. Was everything saved? No, because not everything was able to be regenerated, not everything was able to grow, but a lot of it was." (Scientist, FAO, 6 March 2020: l. 861–868)

The example of the loss and restoration of ICARDA's Syrian seed collection shows that the process of re-establishing the collection in partner institutes in Morocco and Lebanon that followed was more than a process of re-stocking the shelves in those genebanks. Ever since the duplicates stored in the Seed Vault were retrieved from Svalbard, ICARDA's genebank staff has been regenerating and multiplying as many of these samples as possible. This process, which aims to establish a new working collection as well as new safety duplicates going back to the Seed Vault, is expected to continue at least until 2030 (see Westengen et al. 2020: 1312–1313, 1315–1316). As the above-quoted interview passage shows, not every sample from the safety duplicate was entirely retrievable because the regeneration process was not successful in all cases. Notwithstanding that, the fact that safety duplicates had been stored in

Svalbard made it possible to save a large part of the plant genetic diversity in the Syrian collection that would otherwise have been irretrievably lost.

Apart from exemplarily elucidating the value of the Seed Vault, the example of the loss and partial restoration of ICARDA's Syrian seed collection also illustrates what a number of social-scientific accounts of *ex situ* conservation practices have highlighted in recent years, namely that seed conservation, in order to be able to successfully revive frozen seeds and make them available for use, is a process that requires a lot of care work (see Chacko 2019a; Laboissière 2019; Boschen 2022). Such care work, in turn, requires a lot of time and funding, well-equipped and well-staffed genebanks, as well as appropriate environmental conditions. This underscores the importance of international cooperation and a legal framework to underpin it, such as provided by the FAO bodies and the Seed Treaty.

Against this background, the case of ICARDA in Syria both corroborates the logic of double safety duplication and highlights the significance of strengthening "robust infrastructures *and* the capacities of professional staff" (Curry 2022b: 666, emph. added). According to the historian of conservation science Helen A. Curry, these aspects have been neglected since the conception of security underpinning seed conservation practices shifted from a notion of security through capacity-building to a conception of security through backup between the 1960s and 1990s. However, as Curry notes, the privileging of making copies as a cost-effective measure of securitisation "has produced only an elusive security" (2022b: 682) insofar as the roots of the threats that backups are supposed to secure against remain unaddressed. The loss of the Syrian seed collection and the experience of being able to re-establish most but not all of the diversity it previously comprised increased awareness within the *ex situ* conservation community of the perils of securing seed collections primarily through backups (Westengen et al. 2020: 1315–1316). While this does not mean that *ex situ* conservationists question the system of *ex situ* conservation as such, in their reflexion on the case of ICARDA in Syria Westengen and colleagues underline the benefits and indispensability of safety duplication and its limits as well as the necessity of infrastructures and capacities to recreate working collections from backups.

A Leaky Promise of Security?

An important point in critical accounts of the Seed Vault and its rationality of security through backup is the fact that the Seed Vault itself is no more immune to the kind of disasters jeopardising collections held in genebanks around the world than those genebanks themselves (see GRAIN 2008; Nadim 2018: 245). This does not necessarily undermine the Seed Vault's promise of security insofar as it rests on a logic of double safety duplication, which implies that the diversity securely stored in Svalbard is supposed to always also exist in at least two other genebank collections: in the original collection and in the first of two backups. What is more, as Fowler has always pointed out, the idea behind the Seed Vault is not to make "absolute guarantees" (2008b: 15) but to provide the *safest possible* backup infrastructure currently conceivable to secure the world's agrobiodiversity against irretrievable losses in the event of a failure of other conservation infrastructures. One of the factors underlying this promise, that the Seed Vault is the safest possible storage space there is, is the Arctic permafrost and the guarantee it is seen to embody: to provide stable sub-zero temperatures securing the seeds inside the vault against a possible failure of the technical refrigeration system (for the other factors see chapter 1.5). In other words, the Arctic cold is the safety backup for the technical refrigeration system cooling the Seed Vault further down to -18 degrees Celsius.

In recent years, though, it has become increasingly evident that it is not only the technical refrigeration system that is vulnerable. The Arctic is one of the regions on earth where the effects of global warming first became palpable and, to this day, give rise to the most drastic measuring data. According to NASA (n.d.), in 2024 the decrease of Arctic sea ice per decade reached a rate of 12.2 per cent compared to its average extent between 1981 and 2010. Warming temperatures are both cause and effect of this, which is why the Arctic is more strongly affected than other parts of the world. Between 1912 and 2011, the yearly average temperature measured in Longyearbyen increased by around 0.25 per cent per decade (Norwegian Polar Institute n.d.). A report on *Climate Change in the Norwegian Arctic* by the Norwegian Polar Institute predicts that towards the end of the twenty-first century, among a number of other issues, global warming will lead to a severe increase in temperature, precipitation, frequency of extreme weather events, and permafrost thaw (Øseth 2011: 26–41).

The Seed Vault became associated with these developments in October 2016 when a critical mass of water flooded the tunnel leading from the concrete entrance of the facility to the seed chambers inside the mountain. This seemed to prove the point critics had been making about the Seed Vault being exposed to local disasters in the same way that other seed banks are. Many of the news headlines addressing this incident showed no doubt about the reason for this water intrusion: "Norway's Underground Doomsday Seed Vault Is Under

Threat From Climate Change" (Time 2018); "'Doomsday' seed vault meant to survive global disasters breached by climate change" (The Verge 2017); or "The Arctic Doomsday Seed Vault Flooded. Thanks, Global Warming" (Wired 2017).[17] This sort of media coverage points out the irony of a climate change-caused disaster jeopardising the Arctic stronghold for the world's agrobiodiversity, which many consider to be a safeguard against disasters caused by climate change itself.

The story as told to me by the people more directly involved with the Seed Vault does not posit global warming as the main culprit in this incident. According to Åsmund Asdal, the coordinator of the Seed Vault, melt water intrusion has been a continuous issue during the snow melting season ever since the vault was built, albeit previously in negligible amounts:

> "It was written in our annual reports every year that there was water intrusion during snow melting times in summer, but the volumes of water were very low. But in October 2016, there were extremely heavy rainfalls here and so the amount of water was much higher than before; and that convinced all partners that something had to be done." (Åsmund Asdal, Seed Vault Coordinator, NordGen, 7 June 2019: l. 805–812)

As Asdal explained it to me, the significant increase in the amount of water intrusion in 2016 compared to previous years can be attributed to two factors. First, in the Arctic, global warming has lately led to more permafrost melting in the summer than in the past as well as to a higher probability of rainfall instead of snowfall. Hence, the volume of water accumulating around the Seed Vault was simply higher than before. The second factor, which according to Asdal is much more relevant, is the original design of the facility. For one thing, the tunnel leading from the entrance element to the seed chambers was water-permeable to begin with. The primary cause for the large amount of intruding water, however, was that when the Seed Vault was built, insufficient attention was paid to how this would alter the environment. The construction of the Seed Vault severely destabilised the permafrost, which did not reestablish around the tunnel as expected. Although the seeds inside the vault were ultimately not jeopardised because the intruding water never reached the storage rooms, the

17 A google search for "Svalbard Global Seed Vault" with a filter for news published between October 2016 and December 2017 shows many similar headlines on the first few pages.

incident prompted the partners behind the Seed Vault to reconstruct the facility.

The reconstruction of the Seed Vault was carried out between 2017 and 2019 under the aegis of Statsbygg, the company responsible for Norwegian government estates. After Statsbygg had turned down my initial interview request, an unanticipated opportunity arose later in the research process to conduct an interview with one of the engineers who had worked on the reconstruction site. Their account of the water intrusion and the measures taken to secure the facility enriches and differentiates the story as told by Asdal. According to the engineer, the problems with the original design and construction of the Seed Vault that allowed for a critical mass of water to intrude in 2016 in the first place go back to a lack of knowledge about permafrost and to unforeseen changes in the Arctic climate and weather in the years after the facility was built. As they explained the situation to me, the mountains in Svalbard are lithic at their core and covered by a thick layer of permafrost on the surface. While permafrost is generally assumed "to be a mountain, [...] it's really just a lot of water and sand" (Engineer, Statsbygg, 26 March 2023: l. 80–81). When the Seed Vault was first constructed, parts of the surface layer were removed from the mountain it was built into and then raised again on top of the tunnel. The engineers who built the Seed Vault expected that the permafrost would reestablish itself in the Arctic cold, and thus form a natural watertight layer around the steel tunnel leading down into the seed chambers built into the lithic part of the mountain. However, the permafrost did not reestablish itself as expected; as the interviewee trenchantly put it: "Permafrost is an angry lover. You touch her the wrong way, she won't come back." (Engineer, Statsbygg, personal communication) Instead of returning to a permafrost state, the mass that remained on top of the tunnel into the Seed Vault after its construction and before its reconstruction was a porous layer of sand that the heavy rainfalls in 2016 simply "washed away" (Engineer, Statsbygg, 26 March 2023: l. 82). Crucially, these rainfalls were equally unexpected because up to that point, according to current knowledge, "Svalbard has been what they call an Arctic desert" (Engineer, Statsbygg, 26 March 2023: l. 87–88), that is, an environment where there is little to no precipitation. Hence, "in part, it is fair to say that it [the water intrusion; FV] has something to do with climate change. But at the same time, if the permafrost had reestablished itself, it probably would not have been a problem. But of course, we [will] never know." (Engineer, Statsbygg, 26 March 2023: l. 115–119)

The engineer's account indicates that the original design and construction of the Seed Vault as the secure facility it was envisioned to be was influenced by two types of limits to knowledge arising from different kinds of material bounds to knowledge. First, ecological changes on intersecting global and local scales, which have accelerated in recent decades due to global climate change, continuously create new, unknown and often unpredictable environmental conditions that challenge established knowledges, as the example of climate and weather changes and their consequences in Svalbard illustrates. Second, the misjudgement regarding the reestablishment of the permafrost after the construction of the Seed Vault is, at its core, a consequence of the fact that knowledge, to put it with Haraway (1988), is inevitably situated and partial:

> "[Permafrost] doesn't really [come back; FV] if you mess with it. But I guess, that was a knowledge that was not quite known when they constructed the first time. Or maybe it's a very particular knowledge. I mean, you can be a very good engineer, but if you come from mainland Norway, this is not something you know. And in Svalbard, people change all the time. It's just a working environment, so most people go home after a few years. Some stay, but not many engineers have stayed. So, there are very few experts on this." (Engineer, Statsbygg, 26 March 2023: l. 91–101)

Statsbygg's engineers developed the measures taken for the Seed Vault's reconstruction after the water intrusion based on the recognition of this previous lack of knowledge and by bringing in the required expertise. Three main measures were taken. First, the tunnel was reinforced and made watertight through several layers of different materials. Second, a technical room previously located inside the Seed Vault was moved to a new outside building to protect the technical equipment from jeopardy in case of another (if now unexpected) water intrusion. The third and most important measure, according to the interviewee, was the reconstruction of the permafrost. As they told me, by drilling holes deep into the surface layer of the mountain and filling them with cold water, "we reestablished the permafrost with years of freezing down the earth" (Engineer, Statsbygg, 26 March 2023: l. 123–124). The result of these reconstruction measures, as the engineer puts it, is "double and triple and quadruple security for the future" (Engineer, Statsbygg, 26 March 2023: l. 120–121). What remains are the limits to knowledge, not all of which are known and which usually only become discernible when infrastructures fail, as the water intrusion incident proves.

The Politics of the Seed Vault

The engineer's story of the Seed Vault's reconstruction after the water intrusion incident seems to lend plausibility to a risk-oriented framework of analysis, such as developed by Ulrich Beck (1992, 1999) in his theory of the (world) risk society. In view of various large-scale techno-ecological catastrophes that occurred in the second half of the twentieth century and made the hazards of technological progress tangible, like the chemical and nuclear disasters of Bhopal (1984) and Chernobyl (1986), Beck argued that modernity's dichotomic understanding of society and nature had become destabilised.[18] According to Beck, the awareness they triggered of society's intrinsic dependency on an industrially integrated nature heralded the start of a second, reflexive modernity. In this new reflexive stage of modernity, it is no longer the idea of a bright future of technological progress that determines present action, but a threatening if yet unreal future of (self-made) technical risks with destructive ecological ramifications. While sociological analyses of risks and risk discourses made important contributions to the sociological analysis of the interplay of society, nature, and technology, critics, especially in environmental sociology, have pointed out the limits and selectivity of the risk terminology and its insufficiency for a comprehensive analysis of many of the most perilous environmental problems of the present. One important aspect of this critique, raised by Peter Wehling (2011: 533–536; see also Beck/Wehling 2012), is that by focusing on future challenges and how to avert or handle them, risk-oriented perspectives fail to take into account the historical social-ecological contexts from which these risks emerge in the first place.

While the stories of the Seed Vault assembled so far all seem to be set in a world full of risks, variously evaluating the attempts to guarantee security (of seeds, infrastructures, or social-ecological futures) in the face of these risks, the reception of the Seed Vault in parts of the non-governmental sector – similar to Curry's critical analysis of conceptions of security unfolded above – resonates with the sociological critique of the risk-oriented approach. Henk Hobbelink, the co-founder and until recently the coordinator of GRAIN, an NGO engaged in seed politics since the early 1980s, elaborates his perspective on the Seed Vault as follows:

---

18 Another pertinent example is the southern corn leaf blight of the 1960s and 70s, which globally destroyed large amounts of the yields in monoculturally farmed fields cultivated with new hybrid corn varieties (see chapter 1.2).

> "When Svalbard [the Seed Vault; FV] was created [...] our worry was – and still is – its timing, its whole work in the media, the way it is being pictured, being projected as the solution. It gives a false sense of security, as if we have the problem solved. We are not against Svalbard in principle, but against the way it is being used [...] to suggest that by creating genebanks you can solve the problem of the destruction of biological diversity. That is obviously not the case. That is our main issue with it." (Henk Hobbelink, agronomist, GRAIN, 1 May 2020: l. 636–648; see also GRAIN 2008)

What Hobbelink calls into question here is the idea of a simple, technological solution for a problem or set of problems that is much bigger than any technology could "fix" (see also Ribeiro/Shand 2008). He criticises the Seed Vault not for what it is or does in itself, but rather for the techno-solutionist (mis)understanding of agrobiodiversity conservation that it embodies and perpetuates. By iconically materialising the heroic promise of sustaining the world's agrobiodiversity through seed banking, Hobbelink argues, the Seed Vault makes it more difficult for other measures against ongoing agrobiodiversity loss to gain more approval or even attention.

Hobbelink also argues that the idea of seed banking as a technological fix for the loss of plant genetic diversity diverts attention from the forces at the root of "the destruction of biological diversity", which collecting and conserving seed diversity, no matter how securely, will not solve.[19] His choice of words here is noteworthy: Defining the *destruction* of (agro)biodiversity, rather than its *loss*, as the problem (re)politicises the issue. From this point of view, the status quo that the Seed Vault preserves is not only that of an endangered agricultural biodiversity, but also that of an agriculture that puts biodiversity in danger in the first place. Accordingly, rather than criticising *ex situ* conservation as a measure against agrobiodiversity loss in general and assessing the adequacy of the Seed Vault's contribution to it, Hobbelink problematises the effect that

---

19 The anthropologist Tahani Nadim similarly argues that "a project such as the Doomsday Vault [...] not only fails practically but more worryingly perpetuates a logic of ruination" (2018: 245), insofar as the seeds it preserves far into the future will not fulfil their purpose if that future no longer has fields, soils, or people and practices to tend them (see also Folkers 2017: 369–370). While she certainly makes an important point about what might be more broadly referred to as putting life on ice (Kowal/Radin 2017; Radin 2017) or the suspension of life (Lemke 2021), with regard to the Seed Vault this criticism hinges on a narrow understanding of a "doomsday vault" assumed to fulfil its purpose in a faraway future.

the hegemonic power of *ex situ* conservation has within international seed and agricultural politics and the ways in which the Seed Vault, or rather the mis/understanding of agrobiodiversity loss and conservation it communicates to the public, stabilises these power relations.

A slightly different position that challenges this criticism to a certain degree has been articulated by Pat Roy Mooney. Mooney is a civil society activist and influential critic in the field of biotechnology and biodiversity politics, as well as a former associate of Cary Fowler.[20] In the interview I had the chance to conduct with Mooney, he advanced the following position on the Seed Vault:

> "I think it is fine – which is a shock to everybody, of course, because I know that civil society is not supposed to like it. I have always felt that that was based upon a lot of sloppy thinking and sloppy research and misinformation. I don't think Svalbard is the solution to the world's problems – nor does anybody else, I think. But for its cost and for the fact that it is a backup to backup systems – so, it is the court of last resort for seeds – given the price of it, it is a very practical, reasonable thing to do." (Pat R. Mooney, Civil Society Activist, ETC Group, 6 May 2020: l. 141–152)

For Mooney the Seed Vault is, first and foremost, a cost-effective answer to the need for safety duplication and therefore a "practical, reasonable" solution – not "to the world's problems" at large, but to a very specific need in global conservation efforts. He goes on to elaborate:

> "*Ex situ* [conservation; FV] has some value, that is why Svalbard has some value. I do think the future is going to depend upon what farmers save for themselves and around the world and that that is most important. The second-level importance is what is stored at national levels by governments, and then moving up. So, in a way, the least important is what is in Svalbard and second least important is what is in the regional genebanks of the world or the CGIAR centres and so on. But still: I am not against them or anything, I

20 Mooney was one of the earliest activists who intervened in international seed politics beginning in the 1970s (see e.g. Mooney 1979, 1983). In the early days of their activism he worked closely with Fowler, among other things as co-authors of *Shattering* (Fowler/Mooney 1990) and as co-founders of the Rural Advancement Fund International (RAFI) together with Hope Shand. The latter, which has operated under the name ETC Group since 2001, was one of the first CSOs in the field of agricultural and seed politics and remains one of the most influential ones to this day. For more information, see https://www.etcgroup.org/ (last accessed July 18, 2025).

> know they are not made so that the world thinks that the problem has been solved because we've got Svalbard. I don't find that to be the case. I know we say that to each other in civil society – I don't think it is true, I don't think people believe it is true, I don't see governments acting like it is true. What I do see happening is that we've got big seed companies going around the world duplicating everything that is in genebanks and not really concerned about what is left in genebanks after they have made their duplications." (Pat R. Mooney, Civil Society Activist, ETC Group, 6 May 2020: l. 179–199)

Mooney makes an important distinction in this passage between different levels and actors in global seed conservation efforts and seed politics that is crucial for a critical analysis of the latter. Following his line of argument, the Seed Vault might be used and thus appear as an end in itself on the supranational and global level, more specifically if one looks at the interests of the international seed industry. On the national and subnational level, however, according to Mooney, governmental efforts to address the agrobiodiversity crisis do not suggest that the problem is considered solved thanks to the fact that national and sub-national seed collections are safely duplicated in the Seed Vault (if they are). National and intergovernmental agricultural politics and the activities of the international seed industry aside, what becomes discernible in Mooney's critical stance towards arguments articulated by civil society actors such as GRAIN or Hobbelink is that, in his view, they are mistakenly directed at the Seed Vault. The Seed Vault has generated an unprecedented amount of public attention to seed banking that, according to Mooney, serves the cause of crop conservation beyond its own efforts: "It has drawn, I think, a lot of international attention to the importance of genetic diversity, of crops. Probably nothing has caught more attention than Svalbard in terms of this issue. [...] This being discussed is a good thing." (Pat R. Mooney, Civil Society Activist, ETC Group, 6 May 2020: l. 243–251)

This perspective came up more than once in the interviews I conducted for this study. The following passages are from an interview with a plant scientist who has been working for Bioversity International (formerly IPGRI, IBPGR) for a long time, that is to say for an organisation with one of the longest histories in agrobiodiversity conservation and co-initiator of the first Svalbard Initiative together with the FAO. According to this interviewee, generating international attention to the importance of plant genetic diversity, the dangers of its loss, and the resulting need for conservation was not only a welcome side effect of establishing the Svalbard Global Seed Vault but one of the driving forces behind it:

> "I think most of its value is raising awareness. It's a real public awareness story and that has a lot of value. You cannot do that with a genebank, a boring genebank in a country where it is just an old crumbling building or a couple of deep freezers; that doesn't excite the public. But the story behind Svalbard has great value and that is what was also invested in by the people who were the visionaries: the public awareness impact that this would have on conservation. I think that is a very important component of the of the global conservation system." (Scientist, Bioversity International, 20 January 2021: l. 563–574)

This perspective adds another layer to the Seed Vault's cost-benefit-efficiency Mooney invokes in the passage quoted above, by defining public awareness of the reasoning behind conservation measures as a value in itself. Importantly, this is a value that the Seed Vault is able to generate in a way other genebanks are not, as the interviewee sets out in more detail:

> "Nobody wants to fund a genebank, that's a reality. It's like nobody wants to fund a library, it's not exciting. And a genebank is a live library. And so, the funding strategy had to shift because these collections were just being very neglected, they were being funded by small project money here and there, and their long-term funding was not secured. [...] So that is how this initiative [the Seed Vault; FV] came to be realised. That – *ex situ* conservation – is probably one of the hardest things to get donor funding for." (Scientist, Bioversity International, 20 January 2021: l. 594–611)

The arguments that both the scientist from Bioversity International and Mooney put forward offer some more nuance regarding the ways in which the Seed Vault is an important part of the global system of agrobiodiversity conservation beyond providing a secure backup space for seed collections. While this might create the impression of a generally more differentiated understanding of the Seed Vault than articulated by Hobbelink or GRAIN, the two perspectives differ not so much in accuracy as in how they hold the Seed Vault accountable politically. While GRAIN problematises the Seed Vault as a strong symbol of the wrong *answer* to the agrobiodiversity crisis, Mooney highlights the way in which it advances the right *questions*. Put differently, in Mooney's view and that articulated by the scientist from Bioversity International, the Seed Vault draws attention to the *problem* of agrobiodiversity loss or destruction, whereas for GRAIN it draws attention to a questionable *solution* to it. In the following, I unfold how these different ways of looking at the Seed Vault illustrate two divergent notions of the political function or impetus attributed to the Seed Vault and the different actors behind and around it.

Although Hobbelink does not explicitly declare the Seed Vault a techno-fix, he criticises it for diverting attention from the destructive forces underlying the agrobiodiversity crisis. This argument ultimately amounts to the same thing insofar as it suggests that as a technological approach to a problem that actually requires political attention, the Seed Vault is an inadequate solution. Inherent in this criticism is a certain paradox regarding the understanding of the political it articulates. On the one hand, it separates the technological from the political in a way that does not account for the ways in which techno-solutionism *is* a political response to a given problem. Instead, it assumes that the political happens somewhere else than science and technology, thus reproducing the split between politics and science that has given rise to the idea of such a thing as a techno-fix in the first place.[21]

On the other hand, the argument that the Seed Vault creates "a false sense of security" (see above) through its presence in the world and due to the promises associated with it does credit it with a certain political power after all. This power lies in the Seed Vault's assumed capacity to paralyse political action, in seed politics as well as in agricultural politics more generally. This story imagines the Seed Vault as a technocratic intruder in the world of agricultural politics, obstructing 'genuine' politics with its technological promise of security. It allows for differentiating the Seed Vault into what it is designed to be (originally by its creators and now by its managers) and what it is perceived to be (by a vague public or by the undifferentiated media through which this assumed public comes to know the Seed Vault). However, the political agency ascribed to the Seed Vault seems to be anchored in its ascribed status as a techno-fix more than in its effects in the political arenas in which it assumes relevance. Such an interpretation of the Seed Vault as a self-contained and stable entity and agency in the world of agricultural politics

21 As I have indicated in the introduction to this chapter, with Haraway (1997), Latour (1993, 2004a, 2013, 2017), and other STS scholars, my aim in this inquiry into the Seed Vault and the world of agrobiodiversity conservation is to develop a critique that goes beyond such dualisms.

leaves little room for the Seed Vault to be or become something other than what it was designed to be – for example through the relations it enters into.

The understanding of the Seed Vault's political effects articulated by Mooney and the scientist from Bioversity International, in contrast, counters this interpretation with one that leaves much more open what the Seed Vault's political effects are and might become. Here, the Seed Vault does not so much close the case of agrobiodiversity loss and conservation as it discloses it to the world. This makes it possible to focus attention on how the Seed Vault, by drawing unprecedented attention to the issue of agrobiodiversity loss, creates an arena, in which the what and the how of addressing this issue become a subject of collective negotiation. Rather than as a fixed and stable technology intervening in or even obstructing seed and agricultural politics, then, the Seed Vault appears as an agency or a thing that assembles political actors and processes.

This way of thinking about the Seed Vault finds a possible expression in Bruno Latour's notion of "*res publica*, the public *thing*" (2013: 337, emph. in orig.). In *An Inquiry into Modes of Existence: An Anthropology of the Moderns* (2013) Latour unfolds an object-oriented conception of politics that conceives politics as a process and praxis in which a thing of public concern assembles a more-than-human collective. What is important to Latour is the active agency of the thing that assembles "a public to gather around it" (2013: 337). The political collective, which equally encompasses both humans and more-than-humans, emerges in the processual, circular movement of assembling around a public thing and mutates co-constitutively with the thing assembling it.[22] In consequence of this understanding, Latour conceives "the political(ly)" (2013: 339) as an adverb rather than an adjective. This means that things are not political per se, but that the political is a constant doing and becoming in collectivity. With regard to the Seed Vault, this means that it is neither political nor apolitical in itself. Rather, the politics of the Seed Vault come to matter in what it does and what is done with it.

---

22 This understanding of politics as a process of assembling an indeterminate and constantly evolving collective echoes the notion of "the assemblage", as used by Tsing (2015), for example. According to Tsing, the concept serves to evade essentialist or fixed notions of collectives, insofar as "[a]ssemblages are open-ended gatherings. They allow us to ask about communal effects without assuming them. They show us potential histories in the making." (Tsing 2015: 22–23)

An important implication of this conception of the political(ly) is that it makes discernible how political processes, practices, and their material effects are mutable and constantly in the making, and so to a certain degree indeterminate:

> "The same movement of enveloping, encircling, embracing, gathering in can thus serve [...] either to fabricate inclusion [...] or else fabricate exclusion. And nothing in this movement ensures its duration [...]. Everything depends on its renewal, on the courage of those who, all along the chain, agree to behave in such a way that their behaviour *leads* to the next part of the curve." (Latour 2013: 342, emph. in orig.)

Conceiving the Seed Vault in this Latourian sense as a *res publica*, a public thing that assembles an indeterminate political collective,[23] makes it possible to take seriously the "false sense of security" it evokes for some while at the same time understanding it as not merely an inhibition to politics, but also a call to political action. While the Seed Vault certainly is a technology or set of technologies and practices for *ex situ* conservation that has emerged as part of the global system of *ex situ* agrobiodiversity conservation, at the same time it is more than that. The Seed Vault does not determine the collective it calls to action – *ex situ* conservationists only, for instance – and thus prefigure the kind of action that will be taken. Rather, the collective that assembles in the arena the Seed Vault creates and that collectively shapes the next curves of the political turns the Seed Vault takes and may take is an effect of the diversity of responses to this call to action. Understanding the Seed Vault in this way as a thing, or in the terminology of Situational Analysis as an arena assembling a diverse collective of actors and the worlds they come with, rather than as the technology of a specific world, makes it possible to draw attention to the less evident and less expected actors and worlds that form part of the political collective assembling around the Seed Vault. This is not meant to contest the idea that there are material and epistemic power relations within which some actors assume more

23 In a Latourian frame of thought, it stands to reason that the Seed Vault assembles a more-than-human political collective both outside and inside the seed chambers. Such a perspective exceeds the scope of this study, not least for empirical reasons. As explained in chapter 2.3, different practical research restrictions made it impossible for me to develop a more thoroughly more-than-human research approach, such as through a multispecies ethnography attending more closely to the seeds conserved inside the Seed Vault or in other genebanks.

relevance and influence than others, and that these must be part of any comprehensive analysis. Rather, the point is that an analytical focus on the Seed Vault as a political arena makes it possible to recognise spaces of indeterminacy, unexpected agency, and transformative potential in the worlding practices the Seed Vault enacts, facilitates, and may yet enact.

4. Assembling World(ing)s of Conservation

> Like all technoscientific facts, laws, and objects, seeds only travel with their apparatus of production and sustenance. [...] Seeds are brought into being by, and carry along with themselves wherever they go, specific ways of life [...].
> (Haraway 1997: 89)

The Svalbard Global Seed Vault is home to one of the largest collections of seeds from all over the world. These seeds are materialised carriers of a diversity of ways of life – of worlds and modes of worlding the Seed Vault assembles along with seeds. The previous chapter introduced this argument in two steps. First, by situating the Seed Vault within the history of its home, Svalbard, I explored the continuities between the appropriative frontierism and extractivism that have made the archipelago what it is and the *ex situ* conservationism that the Seed Vault embodies. Reading the history of the archipelago's colonisation since the age of European empire-making as an example of a specific type of world-making enabled me to show that the effort to preserve the largest possible diversity of natural resources is a materialisation of the same mode of worlding that has made nature the reservoir of resources the modern world assumes it to be. In this sense, the Seed Vault ultimately serves to preserve the ontological status of the natural/ised world as a resource for human purposes.

In the second part of the foregoing chapter, I argued that the repercussions of this resourcist relation to the natural/ised world materialise both in the global social-ecological problem of agrobiodiversity loss and in other, local challenges the Seed Vault faces. A pertinent example of this is the story of how its construction in Svalbard's mountainscape changed the nature of the Arctic environment in a way that made the permafrost a porous rather than permanent cover for the vault and caused it to leak into the facility. Insofar as the response to this incident was informed by the same logic of securitisation guiding the Seed Vault's response to agrobiodiversity loss, it seems reasonable to argue that the Seed Vault is characterised by an all-encompassing logic of technological fixing. However, I have argued that it is important to consider the ways in which it is not determined by this logic. Attending to the Seed Vault as a political arena that assembles a diversity of worlds and modes of worlding carried by the seeds it conserves makes it possible to develop a more nuanced understanding of the technoscientific politics of the Seed Vault.

The present chapter develops this understanding further by focusing analytical attention on two divergent, yet interrelated worlds and modes of worlding assembled by and co-constituting the arena that is the Seed Vault. Through the analytical framework of worlds and modes of worlding, it becomes possible to develop an understanding of agrobiodiversity conservation as more than a variable set of practices and technologies. What I focus on in the following, then, is not *how* agrobiodiversity conservation is performed so much as *to what end, in what context, to supply and conserve what world*.

I begin by introducing what I call, following one of my interview partners, the *ex situ* world of conservation. This notion denotes the global system of plant genetic diversity conservation as well as the specific way of knowing, being in, and enacting the world that said system is a manifestation of. I argue that it is as such that it must be considered vis-à-vis other forms of conservation situated in other cosmological realities. This is the aim of the second section, which focuses attention on the radically different more-than-human ways of life embodied in the first Indigenous seeds deposited in the Svalbard Global Seed Vault in 2015 by the Potato Park in the Peruvian Andes. The story of the Potato Park shows that the modern way of knowing and being in the world embedding the *ex situ* world of conservation is not as universal as the universalist cosmology underlying it assumes. It is a living example of the fact that other cosmological and worldly realities exist in this world and have persisted throughout extinctionist colonial histories. Inasmuch as the global social-ecological catastrophes and transformations of the present are rooted in the ecologically destructive resourcist, appropriative, and extractivist modern mode of nature- and world-making, there is much to learn from such other worlds that have experienced and survived world-changing catastrophes and transformations.

4.1 The *ex situ* World of Conservation: Extending the Present of "Life Itself"

> "When I think of genetic resources and *ex situ* conservation, I don't think of Svalbard, because Svalbard is a vault. It has been shown to be very useful, particularly for ICARDA, for Syria. It is definitely a success story. And it is an important ... structure or I don't even know what to call it ... initiative ... it's a very important part of the *ex situ* world. But it's a vault. It is not managed, it is not characterised, it has not done anything but kept – and kept as safe as possible – and then you can ask for it back if you need it [the deposit; FV], but the genebanks are where the work is done." (Scientist, FAO, 6 March 2020: l. 325–336)

The practice of conserving agricultural biodiversity encompasses much more than the storage of seeds in a vault, as this passage from an interview with a scientist working for the FAO indicates. I explained in the previous chapter that the Svalbard Global Seed Vault acts as a backup infrastructure for duplicates of seed collections held in genebanks across the world. As such, it is "a very important part" but not the quintessence of the global system of *ex situ* agrobiodiversity conservation. While it has become a famous icon of agrobiodiversity conservation in the past 16 years, the Seed Vault is only the tip of the iceberg that is "the *ex situ* world" of conservation, as the FAO scientist puts it in the passage quoted above. This ad-hoc phrasing resonates strikingly with Haraway's observation that "[n]othing comes without its world" (1997: 37; see chapter 2.2 and 3) as well as with Situational Analysis' focus on technologies as parts of worlds, both of which inform my analytical approach to the Seed Vault and the world of agrobiodiversity conservation. Following the claim these approaches put forward, that a comprehensive analysis of technoscientific projects requires investigating the worlds in which they are situated, the present chapter extends the focus beyond the Seed Vault to explore "the *ex situ* world" of agrobiodiversity conservation it acts as a backup for.

I begin by introducing how the FAO defines *ex situ* conservation and distinguishes it from other approaches to the conservation of plant (genetic) diversity. Through the interviews I conducted with representatives of various segments of the *ex situ* world, I show that the historical opposition of *ex situ* and *in situ* approaches outlined in chapter 1 has given way to a widely shared complementary understanding of different conservation approaches, according to which each comes with its own advantages. Underlying the idea of complementarity, as I show, are divergent understandings of the life that conservation practices target. Whereas *ex situ* conservation makes plant genetic diversity available as "life itself" (Haraway 1997; Franklin 2000; Rose 2007), *in situ* conservation facilitates the evolution and adaptation of "life beyond itself" (Folkers 2017). However, extending the focus from the divergent purposes and premises of different conservation approaches to the worlds they come with makes it possible to discern that both *in situ* and *ex situ* conservation practices can be part of the *ex situ* world if and when their modus operandi disentangles the life they aim to conserve from the (more-than-nonhuman) environments or worlds in which this life comes to matter. Insofar as in doing so, the *ex situ* world of conservation overlooks how specific worlds and modes of worlding are implicated in the loss and destruction of agrobiodiversity, its promise of "extending the present" (Wolff 2020; Lemke 2021) of endangered crop diversity also applies to the world that has endangered this diversity in the first place.

The Complementarity of *ex situ* and *in situ* Conservation

The International Treaty for Plant Genetic Resources for Food and Agriculture (ITPGRFA or Seed Treaty, see chapter 1.4) defines *ex situ* conservation as "the conservation of plant genetic resources for food and agriculture outside their natural habitat" (FAO 2001: 2). It differentiates *ex situ* from *in situ* approaches to conservation, defining the latter as "the conservation of ecosystems and natural habitats and the maintenance and recovery of viable populations of species in their natural surroundings and, in the case of domesticated or cultivated plant species, in the surroundings where they have developed their distinctive properties" (FAO 2001: 2). The most fundamental difference between *in situ* and *ex situ* approaches to conservation, according to this definition, is that the former conserve plants within the ecological environments in which they occur, either naturally or as a consequence of long-term cultivation, whereas the latter extract plant materials from such environments and conserve them within technological (genebanks) or techno-ecological environments (field genebanks; see below). Both types of conservation involve their own distinct sets of practices to maintain the viability of the plants in their custody.

According to the FAO scientist quoted above, the choice as to whether a plant is conserved *in situ* or *ex situ* depends first and foremost on the plant itself, that is to say on what characteristics it has and which type of conservational care it requires:

> "The conservation of plant genetic resources [...] includes *in situ* conservation of crop wild relatives and wild food plants in natural habitats; [...] on-farm management for farmers' varieties and landraces; *ex situ* conservation of orthodox seed in seed banks; field genebanks for the conservation of species that either have recalcitrant seeds or are vegetatively propagated; and *in vitro* conservation and cryopreservation." (Scientist, FAO, 6 March 2020: l. 268–278; see also table 2, completed based on FAO 2014, 2019)

Table 2: Conservation Strategies (the author, 2025)

<table>
<tr><th>Conservation strategy</th><th>Conserved Material</th><th>Backup options</th></tr>
<tr><td>In situ conservation in natural habitats</td><td>Crop wild relatives, wild food plants</td><td rowspan="2">Ex situ conservation</td></tr>
<tr><td>On-farm management</td><td>Farmers' varieties/landraces</td></tr>
<tr><td>Ex situ conservation in field genebanks</td><td rowspan="2">Non-orthodox/recalcitrant seeds, plant tissue of vegetatively propagated species</td><td>In vitro and cryopreservation</td></tr>
<tr><td>In vitro and cryopreservation</td><td>Safety duplication</td></tr>
<tr><td>Ex situ conservation in seed banks</td><td>Orthodox seeds</td><td>Safety duplication (e.g. SGSV)</td></tr>
</table>

This passage illustrates that there is a spectrum of conservation strategies exceeding the dichotomic distinction between *in situ* and *ex situ* conservation. What the Seed Treaty defines as *in situ* conservation encompasses *in situ* conservation in the narrower sense, which refers to the conservation of wild plants within their natural surroundings, as well as on-farm management of domesticated and cultivated food plants.[1] The *ex situ* approach to conservation

1 As the table shows, there are different kinds of both wild and domesticated plants that are relevant for *in situ* and on-farm conservation. Wild food plants are non-domesticated plants that are edible just as they grow. Crop wild relatives are non-domesticated relatives of domesticated crop plants that are not edible themselves but are "potential sources of heritable traits for use in crop breeding" (FAO 2019: 11). The term "landraces" denotes cultivated crop varieties "that [have] evolved through many years of farmer-directed selection and that [are] specifically adapted to local conditions" (FAO 2014: 164).

equally comprises various strategies of conservation; the choice between these depends on species-specific requirements.

The FAO scientist distinguishes three different types of plant species: those that produce orthodox seeds, those that produce recalcitrant, also called non-orthodox seeds, and those that propagate vegetatively, that is, produce no seeds at all. The characterisation of seeds as either orthodox or non-orthodox, a distinction coined in the context of seed banking and without meaning beyond it, denotes whether a seed "can [or cannot] be dried to low moisture content and stored at low temperatures without damage to increase seed longevity" (FAO 2014: 165). Seeds that can be dried and frozen for long-term conservation are called orthodox seeds, whereas non-orthodox seeds lose "vigour and viability" (FAO 2014: 165) or die when their moisture content is reduced. Along with vegetatively propagated plants, therefore, non-orthodox seeds require other methods of conservation than orthodox seeds, such as conservation in field genebanks or *in vitro* or cryopreservation.[2] Irrespective of the conservation strategy, the FAO's *Genebank Standards* (2014) commit all genebanks to safety duplication, as explained in chapter 3.2. Safety duplication in the Svalbard Global Seed Vault is only available for orthodox seeds (so far). Therefore, where I go into more detail on *ex situ* conservation practices in the following, the focus is on the conservation of orthodox seeds in seed banks. Where I discuss conservation approaches in a more general sense and distinguish between *ex situ* and *in situ* approaches to conservation, I follow the Seed Treaty's definition cited above, which conflates the various strategies in this broader distinction.

2 *Ex situ* conservation in field genebanks refers to the conservation of plants by way of cultivation. It is a form of *ex situ* conservation insofar as this cultivation takes place in fields created for the purpose of conservation and not in the plants' natural habitats. The conservation of plant tissue *in vitro* "is used for maintenance of plant organs or plantlets in a medium-term time frame (some months up to some years) under non-injurious, growth-limiting conditions" (FAO 2014: 134). It is not suitable for long-term conservation due to slow growth of the material, a continuous possibility of contamination or hyperhydration, as well as the need for constant monitoring. *In vitro* conservation is often, though not necessarily, conducted in preparation or post-processing of cryopreservation. Cryopreservation denotes "[t]he storage of plant organs in liquid nitrogen (-196°C) or above, in its vapour phase (maximum -140°C). In the genebank context, it is used for buds, shoot tips, other meristematic and embryogenic tissue, explants from recalcitrant and (in special cases) entire orthodox seeds, pollen and somatic embryos" (FAO 2014: 162).

As already mentioned, conserving orthodox seeds for the long term, as in the Svalbard Global Seed Vault, requires drying and storing them at cold temperatures. The first of the two passages quoted from the interview with the FAO scientist above indicates that *ex situ* conservation is not limited to drying and freezing. The process of conserving orthodox seeds in genebanks involves a series of steps standardised by the FAO's *Genebank Standards*. These include "acquisition of germplasm, seed drying and storage, viability monitoring, regeneration, characterization, evaluation, documentation, distribution, [and] safety duplication" (FAO 2014: 5, 17–64). In addition to these procedures, there are standards for genebank personnel and security. Viability monitoring is essential because the viability of a sample can always decline during conservation, no matter how good the storage conditions. If the viability of a sample drops below 85 per cent of its initial viability, the *Genebank Standards* recommend regeneration, that is, thawing, sowing, growing, and collecting seeds to create a new sample. Regeneration is performed to increase the viability of a variety in conservation as well as to multiply the amount of seeds of a sample in storage (FAO 2014: 31, 36). The characterisation and evaluation of seed samples and the documentation of the whole process including all relevant factors is crucial because seed collections, as Pellegrini and Balatti put it, "would have a quite limited usefulness" (2016: 2759) without the collection of data and knowledge about those seeds. It is through the latter that they become valuable resources for breeders and, if and where they have access to genebank collections, for farmers.[3]

I have indicated in previous chapters that there are different types of genebanks. They differ with respect to the plants they conserve and the conservation strategies they employ (see above) as well as the objective they pursue (see Pellegrini/Balatti 2016). While some genebanks immediately supply farmers, others primarily support research and breeding and yet others, such as the Svalbard Global Seed Vault, focus exclusively on long-term storage for safety duplication. Often (though not always) genebanks that directly assist farmers operate on a community level, whereas genebanks that work with breeders or have their own breeding programmes and those that offer long-term conservation often operate on a national, international, or transnational level. This

---

3 Pellegrini and Balatti argue that genebanks, therefore, "can be thought of as part of a broader process of 'databasing the world', necessary for the development of a knowledge economy which needs huge amounts of standardized and classified data" (Pellegrini/Balatti 2016: 2759–2760, referencing Bowker 2000, 2005).

has much to do with the history of plant breeding, which played a crucial role in the globalisation of agriculture in the second half of the twentieth century (see chapter 1).

The IARCs under the umbrella of the CGIAR, for example, which today hold some of the largest transnational genebank collections worldwide, were established as part of the effort to 'modernise' agricultural production through the breeding and dissemination of 'improved' varieties of local crops in large parts of the Global South in the second half of the twentieth century. Collecting and conserving the genetic diversity previously cultivated in those countries in seed banks was a measure taken to prevent this diversity from going extinct while also making it available for further breeding. However, as the plant scientist from Bioversity International I interviewed explained, "there were no national genebanks at that time in many countries. Some of these international collections in the CGIAR existed or were established around that time but countries did not have *ex situ* conservation strategies at all in those days." (Scientist, Bioversity International, 20 January 2021: l. 103–107) While many countries do have national genebanks today, the creation of such national-level facilities is a slow and ongoing process. It is hampered not least by the fact that genebanks do not have the same financial resources as transnational actors with ties to the seed industry – which applies even more to community seed banks.

Ex situ conservation became the core international strategy to counteract the advancing loss of agricultural biodiversity in the 1960s and 70s at the instigation of the then most relevant inter- and transnational agro-political and agro-industrial actors, first and foremost the FAO and the CGIAR. As shown in chapter 1, historical accounts of this development suggest that *ex situ* conservation prevailed over *in situ* approaches for economic, (geo)political, and technological reasons, all of which were embedded in a larger regime of modernisation (see e.g. Curry 2017; Fenzi/Bonneuil 2016; Flitner 2003; Flitner/Heins 2002). I asked all my interview partners about their expert opinion on this historical development and, in some cases, their personal experience of it. The answer I received from Kent Nnadozie, acting secretary of the Seed Treaty, largely aligns with the historiographical narrative. However, Nnadozie also emphasised that deriving a generally adversarial conception of *ex situ* and *in situ* conservation from this development would be an erroneous conclusion:

> "It was more of a misunderstanding of some of the developments [...]. Somehow that has now evolved into like an entity in itself. [...] There was a major, I would say, friction between *ex situ* and *in situ* conservation because the ad-

> vocates of collection and preservation, who promoted the establishment of the CGIAR centres as part of their role then, were much more politically connected and stronger and had more access. The tendency that a lot of attention and resources were diverted to their efforts [...] is now evolving into a demarcation from [*in situ* efforts]. But to my mind, the people who were involved in those activities did not see such a distinction, this was much more a meta-definition." (Kent Nnadozie, Secretary, Seed Treaty, 26 March 2020: l. 787–809)

According to Nnadozie, at the time that the loss of agricultural biodiversity was recognised as a pressing issue on an international level, advocates of counteracting agrobiodiversity loss by collecting plant genetic resources and conserving them in genebanks were able to attract more public attention and resources (presumably primarily meaning financial resources, but possibly also scientific and biological resources) because of their greater political influence compared to proponents of *in situ* conservation measures. As he sees it, institutions with an explicit focus on *ex situ* conservation and plant breeding, like the CGIAR and its IARCs, emerged and gained influence because they offered a more immediate solution to the perceived urgent need for measures against permanent losses of diversity (see chapter 3.2). Accordingly, for Nnadozie, both the primacy of *ex situ* strategies in international conservation activities and the idea that *in situ* and *ex situ* conservation are inherently adversarial approaches were consequences rather than preconditions of this development.

Conceiving *in situ* and *ex situ* conservationists as proponents of two opposing camps competing for the dominant position in the agrobiodiversity conservation movement, in contrast, fails to see that "they all [all conservation strategies; FV] have specific roles and functions to play in the whole, for both agricultural production and food security. So, drawing that distinction has not been helpful if you ask me." (Kent Nnadozie, Secretary, Seed Treaty, 26 March 2020: l. 760–764) In other words, and in line with the differentiation of conservation strategies unfolded through the FAO scientist's perspective at the beginning of this chapter, *in situ* and *ex situ* conservation are not opposing approaches to the same problem so much as different strategies on a continuum of conservation activities counteracting agrobiodiversity loss.

Nevertheless, Nnadozie admits to a historical "friction" (see above) between advocates of *in situ* and *ex situ* conservation that arose from the differential recognition and international support for the two approaches. However, his account also indicates that today, this friction has given way to the notion that different conservation strategies are complementary. According to Cary Fowler, who is among those of my interviewees who have worked the longest in the field of plant genetic diversity conservation, the idea of a complementarity of *in situ* and *ex situ* conservation strategies first arose in the international agro-political arena in the 1990s, although it did not immediately gain acceptance:

> "I recall in 1995 and 96, when I was at FAO and working on the Global Plan of Action, many people involved in *ex situ* conservation were very critical on *in situ* conservation. And the people who were involved in *in situ* conservation very much wanted what they were doing to be recognised and legitimised; and they started talking about *in situ* and *ex situ* being complementary approaches. I believe that the Global Plan of Action, which I personally drafted, was the first document – official document acknowledged by governments – that described the two as complementary. And that was seen, at that time, by the *in situ* advocates as a big step forward. I think after that, ironically, a lot of the *in situ* people started to attack the *ex situ* [approach; FV], I don't know why; not really consider them so much as a continuum but consider them, ironically, as juxtaposed against each other in the same way that the *ex situ* people had treated the *in situ* people earlier; which I always thought was very strange, that the roles had reversed a little bit. I still think they are complementary, and each has its place, and we just have to realise in a thoughtful way what that place is." (Cary Fowler, 19/20 March 2021: l. 254–280)

This passage sheds a slightly different light on the relation between *in situ* and *ex situ* conservationists. For one thing, according to Fowler's account the adversarial relation between *in situ* and *ex situ* conservationists is much less of an external interpretation than Nnadozie suggests. Fowler rather sees it as a mutable manifestation of historical struggles for recognition and support. In addition, this understanding of the history suggests that the *idea* of a continuum of complementary conservation activities is not the same thing as *practical* cooperation and mutual support. The emergence of the idea of complementarity in the 1990s did not coincide with the emergence of complementarity in practice. However, according to Fowler, writing it into the "Global Plan of Action on the Conservation and Sustainable Utilization of Plant Genetic Resources for Food and Agriculture" (FAO 1996), the key official document stipulating the terms for international efforts in agrobiodiversity conservation after the CBD and before the Seed Treaty, was a crucial step in that direction. It created a policy situation in which complementary conservation projects were encouraged and thus incrementally realised.

Today, the complementary understanding of *in situ* and *ex situ* conservation has by and large become common sense among crop conservationists, as most of my interview partners suggest. This does not mean that there is no ongoing friction between proponents of different approaches to conservation (I come back to this below), but the need to forge links is increasingly recognised and links are increasingly being forged. The FAO scientist quoted above concluded their delineation of the different types of conservation – *in situ*, on-farm, and *ex situ* – by pointing out the ways in which they are complementary:

> "I really see all three types of conservation as very complementary. You cannot have one without the other. [...] This is my view and I believe it is also FAO's view that they are complementary. *Ex situ* conservation is absolutely necessary to make sure that as much of those species that are important for food and agriculture are conserved for as long as possible. The advantage of having them in genebanks, besides the security, is that often there is a series of steps that are taking place; often they are characterised, some of the better ones are evaluated, some of them go into breeding programs to help enhance and improve crop varieties that then go back to the farmers to help them improve their yields and their livelihoods. So, there is a lot of work that is done in genebanks that is very important and useful beyond just safeguarding. On-farm management and *in situ* – and more so the *in situ* conservation – means that they are allowed to evolve; [...] and not just evolve but adapt. [...] Especially the farmers' varieties that are adapting allow farmers to grow something that is resilient. Until the case be that plant breeding catches up. The problem is that crop improvement sometimes causes diversity to decrease and increased diversity uses less improved varieties. So, the idea was to have sort of a compromise. Because farmers need to eat." (Scientist, FAO, 6 March 2020: l. 293–324)

This passage illustrates that agrobiodiversity conservation means different things and serves different purposes. A key goal of *ex situ* conservation is making PGRFA available for the long term. The Svalbard Global Seed Vault, being a storage space for safety duplicates of other genebank collections, serves solely this purpose by way of safeguarding these collections. Other genebanks, such as those depositing their duplicates in Svalbard, work towards long-term availability of PGRFA by way of actively maintaining their collections. While they need safety duplicates to insure their collections against the many kinds of threats they are exposed to, the viability and value of safety duplicates depends on the work done in genebanks.

Beyond guaranteeing long-term availability, *ex situ* conservation also facilitates the production of new crop varieties through breeding by offering a supply of PGRFA and data collected alongside them. To breed 'improved' varieties and supply farmers with these varieties, which will help them achieve better yields, genebanks and breeders need an ongoing supply of a diversity of PGRFA. Therefore, they depend on farmers who cultivate that diversity and on *in situ* conservation efforts that maintain resources of potential value for food and agriculture in the wild. In a nutshell, *ex situ* conservation "is a backup for the farmers, but [...] the farmers are also a backup for the genebanks" (Scientist, FAO, 6 March 2020: l. 128–130). What *in situ* conservation and on-farm management make possible that *ex situ* conservation and breeding do not is the continuous evolution and adaptation of PGRFA within the ecosystems in which plants ultimately will have to grow and thrive. Thus understood, a complementary understanding and practice of different forms of conservation is not a 'compromise' so much as mutually beneficial.

This perspective is similarly expressed by another of my interview partners, Godfrey Mwila, who has been actively involved in the plant genetic resources conservation movement since the 1990s. He has worked in different institutional contexts and capacities, mostly engaged in the establishment of national genebanks in Southern Africa and policy development on the international level. In contrast to the interviewee working for the FAO, who is a scientist with a background in conservation genetics and whose viewpoint is unambiguously situated within the *ex situ* world of conservation, Mwila's perspective is somewhat differently situated. His main concern is strengthening on-farm conservation and sustainable farming systems more generally. While he shares the conviction that "genebanks should be seen as a complementary system for on-farm seed security" (Godfrey Mwila, 12 March 2020: l. 247–249), he makes the case for complementarity, so to speak, from the other side:

> "On-farm conservation activities or programmes have not been that much supported compared to *ex situ* collections; and the on-farm conservation system is important because that is where the materials continue being grown and continue being adapted to the changing environment and so it is important that those systems are also supported. Linkages should create a situation where private actors like seed companies and other donors see the importance of on-farm activities and start supporting them, bring farmers and *ex situ* conservation and also the use part closer together. Because this is a continuum. What is happening on-farm will have an effect on what is hap-

> pening in the business of seed production and supply and also will have an effect on the future functioning of genebanks, of *ex situ* collections." (Godfrey Mwila, 12 March 2020: l. 277–298)

Like the FAO scientist, Mwila emphasises that the adaptation of plants to changing environmental conditions happens *in situ* and through on-farm cultivation. However, while the former argues that these are important sites of conservation "[u]ntil the case be that plant breeding catches up" (see above) by producing improved varieties of crops adapted to local conditions, Mwila calls for more support for *in situ* and on-farm conservation. He argues that linkages between farmers, genebanks, and the seed industry are needed to put complementarity into practice and orient conservation towards use.

Conservation for use, as nearly all my interview partners suggest, is a guiding principle in all forms of conservation. This is reflected, for instance, in the following passage from my interview with Cary Fowler:

> "For me, the purpose of conservation – even though I get good warm feelings from seeing everything conserved – is really to use it and that it is for a purpose, at the end of the day. I think where that maybe intersects with *in situ* is that farmers [...] are conserving genetic diversity, but that's not the purpose of what they are doing. [...] There are very few farmers that are farming for the purpose of conserving diversity. They want to grow food and make a living and that kind of thing." (Cary Fowler, 19/20 March 2021: l. 300–311)

Fowler draws a connection between different types of conservation insofar as they are all ultimately oriented towards use and thus towards a concrete purpose beyond conservation itself. However, it remains opaque what "use" and what "purpose" exactly he has in mind when it comes to genebanks and *ex situ* conservation. While food production might be the purpose of all conservation activities "at the end of the day", seeds travel a long way from the genebank to the field, and I have elaborated above the many steps involved. The purpose of most genebanks not directly assisting farmers – and thus that of most genebanks whose duplicates the Seed Vault holds – is to supply research and breeding. This is a type of use that both differs from and affects the use in farming that Mwila talks about.

For Mwila, use clearly means on-farm food production and direct access to PGRFA for farmers rather than breeding:

> "The central theme that should justify linkages is the actual use, is the need to actually increase the use of materials that are conserved in genebanks. And access to those materials by farmers, for example, has not been that good. Given the effects of climate change, which is reducing the resilience of farmers, there is a need for farmers to get more access or to have knowledge about the materials that are in those genebanks. And so, one of the things about these linkages is to create, to develop the farmers' capacity, to increase their capacity – in terms of understanding the role of *ex situ* collections and their own role and how these two systems are really mutually supportive." (Godfrey Mwila, 12 March 2020: l. 259–276)

Forging linkages between farmers and genebanks, as Mwila elaborates in this passage, has a threefold relevance for farmers. The main goal is for them to be able to use the material conserved in genebanks. Therefore, they need better access to it as well as to the knowledge required to put it to sensible use. Both require what Mwila calls capacity-building in order for farmers to become equal partners in the system of conservation and sustainable use. Linkages are particularly important for farmers, as Mwila argues, in light of the environmental challenges that come with climate change and necessitate new degrees and forms of resilience. According to both Mwila and, even more explicitly, the FAO scientist, resilience can be achieved by adapting cultivars to changing on-farm conditions through ongoing cultivation. However, although they agree in a number of respects, it is clear from the way Mwila and the FAO scientist talk about complementarity what part of the conservation continuum they are engaged in themselves and from which standpoint they make a case for complementarity. While from the FAO scientist's point of view, *in situ* and on-farm conservation are a backup and source of evolving diversity for *ex situ* conservation and breeding, Mwila highlights the role of genebanks as sources of supply for farmers.

In contrast to both of these accounts Hannes Dempewolf, who is a scientist at the Crop Trust, expresses an understanding of complementarity that aligns more with Fowler's above-cited conviction that "each has its place; and we just have to realise in a thoughtful way what that place is". In response to my question about the importance of *ex situ* conservation in comparison to other approaches to conservation, Dempewolf states:

> "You are talking to someone from the Crop Trust and we, of course, are all about *ex situ* conservation. That is our mandate, that is our mission, that we

> live and breathe *ex situ* conservation. That is not to say that we do not recognise the importance of *in situ* conservation and its complementarity. In fact, I am, at the moment, developing a project that is trying to get away from this idea of dividing *ex situ* and *in situ* and think about it as a conservation continuum, which I think makes a lot of more sense; and you get away from these antagonistic wars of our research allocations that are really unhelpful. I think genebanks need to support *in situ* efforts and *in situ* efforts need to work with genebanks. That is something that is bound to happen, it has to happen, and I hope it happens soon. [...] By our mandate we are bound to *ex situ* but we always try to see in which way the work we support can help *in situ* efforts. [...] Having said that, they also have slightly different roles. *In situ*, even when it comes to crop diversity, is a changing system. The diversity continues to adapt in the field, whereas in genebanks it is static – apart from regeneration cycles, but it is relatively static – which means that the material that evolves with farmers changes, it is not per se conserved, because it changes with environmental conditions. Of course, that means that the sort of material you conserve and the conservation dynamics differ, they are just different. We have to recognise that there are advantages and disadvantages for both." (Hannes Dempewolf, Scientist, Crop Trust, 28 May 2019: l. 704–759)

Dempewolf differentiates between *in situ* and *ex situ* conservation not only in terms of which plants are conserved, which procedures used, and which goals pursued. The fundamental difference between the two types of conservation he describes is that the conserved material exists in divergent temporal and environmental conditions. His argument that plant material conserved in genebanks is "relatively static" refers to the fact that cold storage slows down the conserved material's biochemical processes of life. Drying and freezing strip seeds of their capacity to change their current state of being and adapt to the environmental conditions in which they exist. In cold storage, therefore, seeds (ideally) neither germinate nor decay.[4] Plant genetic material conserved

---

4 Biochemical processes do not actually come to a halt through freezing but are merely being slowed down, the more so the lower the temperature (see also Laboissière 2019). In addition to temperature, there are other aspects that also influence the degree to which viability is preserved in storage, such as moisture content and the occurrence of pathogens. In 1984, the Nordic Genebank (the predecessor institution to NordGen) initiated a 100-year experiment to monitor the development of the viability of seeds stored inside the permafrost, at -3 degrees Celsius (the samples were stored inside an abandoned coalmine in Svalbard). Interim results after 35 years show a decline in viability which proves that the naturally occurring cold temperatures in the Arctic are

in situ, in contrast, is embedded in a living environment, within and along with which it evolves and to which it adapts. Accordingly, while the temporality of material conserved *ex situ* is relatively static, the temporality of material conserved *in situ* is dynamic; *ex situ* conservation arrests change, whereas *in situ* conservation embraces change.

The idea that plant material conserved *in situ* is "not per se conserved" is highly instructive insofar as it displays the *ex situ* conservationist understanding of plant genetic material as something that exists in itself, that can be disintegrated from a plant or an ecosystem (and introduced into other environments, whether ecological or technological) without changing the nature of what it is.[5] This *ex situ* world view, as it might be called following the notion of the *ex situ* world, resonates with the fundamental premise of technologies of biobanking. According to the sociologist and STS scholar Thomas Lemke (2012), in biobanking "the focus is entirely on securing the existence of life forms as isolated and identifiable entities that can be stripped of their natural environment, technically processed, preserved, and transferred to other contexts" (2012: 191). This definition concisely encapsulates the rationale behind Dempewolf's argument that "the sort of material you conserve" is "just different". Seeds and other isolated plant material conserved *ex situ* are the bio-objects that technologies of biobanking simultaneously assume and produce. *In situ* conservation practices, in contrast, target plants-within-ecosystems, or ecosystems that sustain the life of certain plants. *In situ* conservation does not aim to conserve specific, identifiable, extractible, and technologically processible plant genetic resources so much as ecological processes and the ongoingness of the life of species relevant for food and agriculture within the (changing) ecologies in which they thrive. Hence, that "the conservation

---

not cold enough for successful long-term conservation (see Asdal et al. 2019). In 2020, NordGen launched a new 100-year experiment in which Seed Vault depositors are collaborating to assess the development of viability of seeds stored inside the Svalbard Global Seed Vault, i.e. at -18 degrees Celsius, in relation to growing, harvesting, and storage conditions (see NordGen 2020).

5 This view more generally informs processes of detaching and accessing entities from ecologies in the *ex situ* world, as a current controversy on digital sequence information (DSI) demonstrates. In this case, it is the digitalised genetic information embodied in a seed that some in the breeding industry treat as separable from the material seed. Based on this interpretation, they seek to obtain free access to the DSI by arguing that benefit-sharing regulations only apply to material resources and not to the information encoded in them (see Nehring 2022).

dynamics differ", as Dempewolf puts it, means that *ex situ* and *in situ* conservation ultimately presume and seek to preserve different things: "life itself" (Franklin 2000; Rose 2007) as opposed to "life beyond itself" (Folkers 2017).

The Politics of Conserving "Life Itself"

The notion of life itself became a pivotal point in social-scientific debates about biopolitcs around the turn of the millennium. It is well known today that the concept of biopolitics goes back to Michel Foucault (1970, 1978, 2003), who introduced it to encapsulate a modern form of governing life through the regulation of individual bodies and populations. In light of the biotechnological developments of the late twentieth century, which shifted the focus of biopolitical control from the life and death of individuals and populations to "life itself" on the molecular level, scholars such as Sarah Franklin (2000) and Nikolas Rose (2007) described the rise of a new form of "molecular biopolitics" (Rose 2007: 17). This form of biopolitics is characterised by what Haraway calls a "genetic fetishism" (1997: 141–148). "[T]he chief actor and point of origin in the drama of life itself – the gene" (1997: 133) is fetishised as an autotelic thing-in-itself – which, however, requires erasing the technoscientific apparatuses of production through which genes come to matter in the first place.

Enabling the new forms of governing life in the genomic era, according to Franklin (2000: 188–191), are three historical shifts in the onto-epistemology of life: the biologisation of nature in the modern era, the geneticisation of biology the second half of the twentieth century, and the consequential instrumentalisation and capitalisation of what has thus become "life itself". At the turn of the millennium, these developments led Franklin to conclude:

> "The twentieth-century transformation of life itself has had the consequence that the grounding or foundational function of nature as a limit or force in itself has become problematic and lost its axiomatic, *a priori* value as a referent or authority, becoming instead a receding horizon. Nature [...] has been antiquated, displaced and superseded, and now it is only a trope – a mere shadow of the referent it used to signify." (Franklin 2000: 190–191)

Today, this conclusion seems hardly tenable. Nature as a limiting force has reappeared on the horizon in the form of multiple ecological disaster scenarios presently unfolding and accelerating all over the world. Anthropogenic climate change and biodiversity loss, above all, are casting ever greater shadows on the modern dreams of human exceptionalism and unrestrained technological emancipation from and mastery over nature.

Like previous (geo)historical situations, the current one too has given rise to new technologies of control that come with new forms of biopower and biopolitics. As I have shown throughout this chapter and the previous ones, the growing realisation that vegetal and other forms of genetic diversity and hence a great stock of capitalizable resources are threatened with extinction both *in situ* and in biobanks has resulted in the emergence of cryobanks, meaning conservation facilities that preserve genetic resources for the long term through the use of cold temperatures, or what are now known as cryotechnologies.[6] Technologies and infrastructures of refrigeration were developed as early as the second half of the nineteenth century. Today they facilitate unprecedented and increasingly refined ways of "controlling, transforming, safeguarding, and enhancing life" (Friedrich 2017: 61; see also Friedrich/Höhne 2016). Through the disentanglement of bits of life – both more-than-human and human – from the "biological, ecological and social interactions [they] originated from" (Lemke 2021: 710) and through their suspension in a state of being neither alive nor dead, cryotechnologies serve to make biomaterial indefinitely available as "biological capital" (Lemke 2021: 707; see also Sunder Rajan 2006).

The government of life that these technologies exert, or make it possible to exert, has been theorised as a form of "cryopower" (Friedrich 2017) and "cryopolitics" (Kowal/Radin 2017) or, as Lemke has recently suggested in a critical revision of the former term, a "politics of suspension" (2021, 2022). According to Lemke, this form of governing life "goes beyond conventional technologies of collecting and storing organic material as it [...] not only assembles and preserves 'cryofacts' (Friedrich 2020, 329) for contemporary usages but also mobilizes and explores purposes 'as yet unknown' (Radin 2017, 55)" (Lemke 2021: 711). An excellent example of this that Lemke (2021: 711–712) himself adduces

6 The prefix *cryo* derives from the Greek word *krýo* (κρύος), which means icy cold or frost (see Radin/Kowal 2017: 4). Cryopreservation in the strict sense of the term refers to the conservation of biomaterials at cryogenic temperatures, that is, -196 degrees Celsius, which is the temperature of liquid nitrogen. Conservation through freezing is also possible and performed at lower sub-zero temperatures, for example at -80 degrees Celsius in dry ice, depending on the biomaterial in question and its behaviour at low temperatures. If not specified otherwise, I use the term cryotechnologies as a signifier for all technologies that freeze biological material at sub-zero temperatures, including the conservation and cold storage of seeds at -18 degrees Celsius.

is the conservation of plant genetic material in genebanks that do not directly assist farmers but conserve it for the medium and long term in order to make it available for various, not predefined types of future use. Insofar as the Seed Vault and the genebanks whose duplicated collections it safeguards aim to collect, conserve, and make available "every trait we might want our crops to have in the future – all the options – [that] can be found in this genetic diversity" (Fowler 2016: 18), what they assemble is biocapital with manifold, yet to be realised purposes (see also Breithoff/Harrison 2020; Dalyan 2021).

The promise of an endless continuation of capitalising 'life itself' in ways yet unknown must be situated in the current geohistorical moment, which is characterised by accumulating and intersecting social-ecological crises that challenge the modernist dream of unlimited economic growth and development through ever-expanding resource exploitation. Underlying the Seed Vault's politics of suspension is the promise that there is a way to reckon with worldwide ecological devastation that does not require suspending "the patterns of global production and consumption that have led to the ecological crisis and environmental destruction" (Lemke 2021: 712). Instead, *ex situ* conservation is a cryotechnological promise to preserve a world on the brink of extinction by way of suspending the life in danger of being lost along with the threat of extinction. The temporal horizon of the politics of suspension enacted by long-term seed conservation facilities, then, is not the future so much as the present, as Lemke (2021: 711) argues, building on Leon Wolff's intriguing analysis of the Svalbard Global Seed Vault's "politics of reversibility" (Wolff 2020). Rather than anticipating and acting on the future by striving to prevent, prepare for, adapt to, or accommodate it, the Seed Vault and the politics of suspension it exemplifies "extend[...] the present" (Wolff 2020: 3; Lemke 2021: 711) by postponing future threats or their irreversibility. Where the attempt to preserve the status quo entails postponing political and social change outside the genebank, which might address the roots rather than the effects of the contemporary social-ecological crises, the politics of suspension become conservative in every sense of the term (Lemke 2021: 712; see also von Verschuer 2021: 60).

This assessment resonates with a critique mentioned in the previous chapter, namely that the Seed Vault and *ex situ* conservation evoke "a false sense of security in a world where the crop diversity present in the farmers' fields continues to be eroded and destroyed at an ever-increasing rate" (GRAIN 2008; see chapter 3.2). Beyond objecting to *ex situ* conservation as a techno-solutionist answer to agrobiodiversity loss, as set out in the previous chapter, this critique also draws attention to the world in which the problem of agrobiodiversity arises. This world is not only a world of endangered agrobiodiversity, but one that puts biodiversity in existential danger in the first place. Agrobiodiversity loss and *ex situ* conservation of agrobiodiversity in the form of PGRFA come from the same world; they are manifestations of the same mode of worlding. Extending the present of that world and mode of worlding through *ex situ* conservation, then, also extends the presence of agrobiodiversity loss while at the same time suspending attention to its roots.

It is important to understand that the point here is not to argue against *ex situ* conservation as such. What I mean to problematise is not the practice or technology of *ex situ* conservation itself, but the *ex situ* world's hegemonic position within international conservation efforts. It is certainly fair to say that the *ex situ* world's politics of suspension works towards extending the present, as Lemke and Wolff do. However, and this is crucial, it is the present of a specific world that is extended: the present of a world made of biofacts that can be *ex situated*, biovalue that can be extracted, biocapital that can be accumulated. This world is the product and a continuously produced materialisation of an *ex situ* mode of worlding that operates by erasing both ecological and technoscientific environments producing and sustaining what is fetishised as life itself. Making the conservation of this world and mode of worlding and the extension of its present the core international strategy to counteract agrobiodiversity loss comes at the expense of other, differently cultivated worlds, presents, and futures. These are more-than-speculative, existing and lived real worlds and presents increasingly forced into a reality of devastating futures-becoming-presents (see the following section). Recognising and accommodating these would mean to not conserve just one world and its present, but to cultivate and conserve what the Zapatistas call "a world in which many worlds fit" (Ejército Zapatista de Liberación Nacional as quoted and translated by de la Cadena/Blaser 2018b: 1).

From "Life Itself" to "Life Beyond Itself" and More-Than-Human Worlds

Against the background of the earlier discussion of different modes of conservation, the question arises of what role *in situ* conservation assumes in this context and in what ways it does or does not open up different paths in and for the world of agrobiodiversity conservation as well as towards "a world of many worlds" (de la Cadena/Blaser 2018a). Based on Dempewolf's differentiation of *ex situ* and *in situ* conservation as fundamentally different approaches regarding their conservation dynamics and the material they target, I have stated that *in situ*, other than *ex situ* conservation, seeks to preserve not life itself but life beyond itself. The notion of "life beyond itself" goes back to Andreas Folkers (2017), who introduced it to account for the environments or the worlds that fall into oblivion in the politics of life itself. Folkers (2017: 367) encapsulates this key argument of his text in the German word *Umweltvergessenheit*. There is no equivalent to this term in the English language, and a literal translation would amount to something like "environmental oblivion". It is important to note, however, that the German word for environment (*Umwelt*) encompasses the word for world (*Welt*). Folkers interlinks both terms in the notion of *(Um)Welt*, which illustrates that the world is more than an environment, or to put it another way that there is more to the environment than the term classically suggests. Bringing the *(Um)Welt* back into the discussion of the politics of life itself, as Folkers does, means situating life not merely in genes, cells, and seeds, but also in the ecologies that make them.

In reference to the work of Gilles Deleuze and Félix Guattari (and in contradistinction to currently flourishing neo-vitalist interpretations of their work), Folkers proposes a notion of ecology that denotes not an assemblage of biological lifeforms and their relations so much as an intricate nexus of living and non-living matter. This approach replaces dualist distinctions between nature and culture or biology and technology with an understanding of vitality as emerging in a dispersed fabric of relations "that exceeds both organic life and life itself" (Folkers 2017: 375, author's translation). Folkers condenses this conception in the notion of "life beyond itself", which serves to emphasise that life is not substantial but relational. In the same way that human bodies become what they are in more-than-human relations (Haraway 2008: 3–4; see introduction), ecologies and their components become what they are through ongoing more-than-nonhuman relations and ways of mutual affecting and being affected. Based on this understanding of life, the environment is not a sphere of its own that is merely a surface or surrounding to the lifeforms inhabiting it. Rather, lifeforms and their respective more-than-nonhuman environments – their worlds *[(Um)Welten]* – always form a "symbiotic complex of relations" (Folkers 2017: 373, author's translation).

The rationale of conserving plants *in situ* and on-farm so that they may evolve with and adapt to the (changing) environments in which they grow illustrates this notion of a politics of life beyond itself. Another passage from the interview with Dempewolf, in which he argues that *ex situ* conservation is better suited for long-term conservation than *in situ* approaches, demonstrates that this process of plant evolution and adaptation refers to more than ecological environments in the narrower sense, namely to more-than-nonhuman worlds or *(Um)Welten*:

> "In terms of long-term conservation, if you talk about conserving material forever, which is what genebanks are for, *ex situ* is probably the preferred choice because you have much more control over the conservation process; whereas in *in situ* conservation, you work with people; and farmers are businessmen, they make decisions based on what makes sense for their conditions, and if certain incentives are missing then they may stop conserving certain materials simply because they have to feed their families; and you cannot be ignorant of that either. So, I think for the long term, *ex situ* conservation is probably something that needs to be highlighted." (Hannes Dempewolf, Scientist, Crop Trust, 28 May 2019: l. 760–773)

In this passage, Dempewolf situates not only the plant life that conservation practices target within the environment of conservation. He also situates the practice of conservation itself within the more-than-nonhuman world it is embedded in. Unlike in *ex situ* conservation, especially in preservationist facilities, where use is a long-term goal and does not necessarily refer to the use of farmers (only), farmers do not conserve to conserve but to be able to continue to cultivate crops and make a living. Accordingly, *in situ* and even more so on-farm conservation depend on environments made up not only of climate, weather, and soil – including all the fungal, bacterial, microbial and other beings constituting it – as well as pests, diseases, and other lifeforms, but also of farmers and their contingent decisions about what (not) to grow and conserve as well as the social, political, and economic relations co-constituting these decisions.

Agrobiodiversity loss, then, may result not only from ecological changes and incidents or from technological and human failure jeopardising *in situ* and *ex situ* conservation infrastructures, the reasons usually invoked to vindicate safety duplication (see chapter 3.2). Loss of agricultural biodiversity can equally be a result of non-cultivation or neglect, quite possibly as a consequence of necessity. The "control over the conservation process" that *ex situ* conservation promises, as Dempewolf argues, then, arises from the disentanglement of seeds not only from their biological and ecological entanglements (or more precisely from as many of these as possible), but also from their more-than-nonhuman worlds [*(Um)Welten*].

What remains unmentioned in Dempewolf's account, however, is that the *ex situ* world itself does not exist isolated from other human and more-than-human worlds either. To produce and maintain PGRFA, it depends on technological and infrastructural as well as political and economic apparatuses far exceeding the confines of the *ex situ* world of conservation. In the interview I conducted with a CSO representative in an Ad Hoc Technical Expert Group (AHTEG) advising the Seed Treaty, they explained how conservation activities are entangled in political and economic issues and what relevance that, in turn, has for the discussion of *ex situ* and *in situ* conservation and their complementarity.

> "It is very important to be practical, but very critical on the roles of the different types of conservation. A lot of agrobiodiversity is being lost sometimes in a year particularly in farmers' communities because they lose their seeds due to climate change, for instance. That element makes *ex situ* conservation strategies very important. But it has to be put in context. The reason why they are not very effective is because for farmers to get access to the seeds – whose value they also have contributed to – is so complicated, so expensive. So, the technical idea of having *ex situ* conservation is not bad, it is actually good and complementary to *in situ* conservation. But that strategy is not isolated from the whole economic and political context and that makes it very difficult for some countries and for some actors to access the material and very easy for others. So, that is also part of the discussion on *in situ* and *ex situ* conservation. And that is also part of the reason why community seed banks are very important. [...] So, I believe that *in situ* conservation deserves more attention, more support to fulfil its purpose and also to make conservation more effective and access to seeds more effective and fairer." (CSO Representative, AHTEG/Seed Treaty, 28 September 2020: l. 554–578)

This passage introduces yet another level of distinction between different conservation approaches by foregrounding the question of use through the question of access. While the interviewee does not question the complementarity of *ex situ* and *in situ* approaches when it comes to conservation, they highlight the unequal accessibility to materials conserved *ex situ* that makes it more difficult for farmers, especially in the Global South, than for other actors, such as the seed industry, to use them.[7] It is against this background that the inter-

7 This argument refers to an ongoing controversy in the governing body and working groups of the Seed Treaty, as well as in the FAO more generally, that revolves around

viewee assesses the effectivity of *ex situ* conservation. For them, it is effective not when as much crop diversity as possible is conserved, but when that diversity is accessible for farmers to cultivate. Although the seed industry does of course supply farmers, according to the interviewee access to their seeds is far more complicated and expensive than access to community seed banks, for instance. Most hybrid crop varieties (that is, varieties developed through breeding) are bred in such a way that they do not produce seeds themselves. As a result, farmers cannot obtain seeds from the plants they cultivate and save them for future use but are forced to buy hybrid seed from breeders every year. The resulting dependence of farmers on the seed industry is particularly problematic because hybrid seed varieties are usually patented, which makes them more expensive than landraces and subjects farmers to the pricing policies of the seed industry. In this context, the interviewee emphasises the advantage of *in situ* conservation approaches: they make access to the conserved material easier for farmers. This is important not only because farmers "have to feed their families", as Dempewolf puts it (see above), but because they are the ones who ultimately cultivate the crop diversity that all forms of conservation seek to conserve. This is a present that the politics of suspension does not extend, although it is essential if the life suspended in genebanks is to come into life in the first place.

It is this oblivion of the worlds sustaining what is fetishised as life itself that Folkers (2017) draws attention to. The differentiation between world and environment is crucial here. What the term "world" in contrast to "environment" illustrates is that the world that makes a seed is a more-than-nonhuman world comprising multispecies, including human, agencies and relations as well as the political, economic, epistemic, and other power relations they are embedded in. If and where *in situ* conservation refers to the conservation of plant genetic material within natural habitats or the ecosystems in which "they have developed their distinctive properties" (FAO 2001: 2; see above) while disregarding the worlds in which these sites exist, it amounts to an *in situ* version

farmers' rights, access and benefit sharing. The most relevant conflicting parties are the seed industry, most of which is located in the industrialised countries of the Global North, and farmers' organisations representing farmers in the Global South with the support of NGOs and CSOs. Although the Seed Treaty has established statutory provisions for farmers' rights as well as access and benefit sharing, these provisions have not been sufficiently implemented, and they are still the subject of heated discussions (see chapter 1.4).

of *ex situ* conservation. In other words, this form of *in situ* conservation is part of the *ex situ* world. The question, then, is not only which present is extended and which form of life conserved, but also which world or mode of worlding. In his discussion of the politics of life itself and life beyond itself, Folkers argues that this is neither a question about the (un)naturalness of environments nor a question of right or wrong. Instead, it is one that asks in which world [*(Um)Welt*] life – plant life but also human life – can thrive and to what degree it can unfold its vitality (Folkers 2017: 377–378). The figure of life itself, according to Folkers, is a form of life deprived of its vitality, meaning its capacity to evolve, adapt, and thrive as well as its means of spontaneous expression and aesthetic potential. This deprivation reaches its (current) climax in the cryotechnological suspension of life in cold storage. Extending the focus from "life itself" to "life beyond itself" helps us to see that what the *ex situ* world's politics of suspension suspend is not only the vitality of seeds, but also the vitality of other (human and more-than-human) worlds and modes of worlding. From this point of view, *ex situ* conservation appears as an attempt – or a technological tool serving the attempt – to retame a nature that is out of control and retain its exploitation.

4.2 From Global to Earthly Interdependence: Thinking Conservation as Mutual Care in More-Than-Human Worlds Through the Potato Park

The first time I really became aware of the plurality of worlds the Svalbard Global Seed Vault assembles and the importance of attending to the relations between them was during my second field research trip to Svalbard in February 2020. The occasion of this visit was an international summit convened by the partners behind the Seed Vault in celebration of its twelfth anniversary and the completion of the constructional upgrades to the facility undertaken in consequence of the water intrusion that had occurred in 2016 (see chapter 3.2).[8] Members of the globally dispersed *ex situ* world of conservation assembled in Longyearbyen, Svalbard, and were joined by media representatives from all over the world. I attended the summit as a participant observer.

8 The event took place only days before the COVID-19 pandemic reached a level of global concern that made in-person gatherings of international communities and any further field research temporarily impossible.

The three-day event began with a reception at the Svalbard Museum in the afternoon of the first day, followed by an information programme about Svalbard and the Global Seed Vault. The second day was the day of the main event: The 2020 Svalbard Seed Summit bearing the title "Genetic Diversity for More Resilient Food Systems". Under the banner of the UN Sustainable Development Goal (SDG) number two (zero hunger by 2030) and against the background of the impending failure to reach SDG target 2.5 (preserving the genetic diversity of crops and livestock and advancing access and benefit sharing by 2020), the summit addressed achievements, current issues, and future challenges of the global endeavour of *ex situ* agrobiodiversity conservation.[9] Towards the end of the summit, a delegation of the UN's SDG Advocates, a fluctuating group of influential people from all over the world who use their respective spheres of influence to raise awareness of the SDGs and the need to take action,[10] joined the meeting. The main purpose of this visit was the release of an Arctic Call to Action on Food Security and Climate Change (SDG Advocates 2020) urging the international community to recognise the importance of the SDGs and show more commitment in pursuing their realisation.

The SDG Advocates' special appearance concluded a series of talks and Q&A sessions, throughout which members of the partner organisations behind the Seed Vault, representatives of depositing genebanks, and other stakeholders had discussed the ways in which conserving plant genetic diversity contributes to global food security, which roles different types of genebanks assume and how to strengthen cooperation among them, as well as the technical and political challenges connected to all of this. A pivotal motive of all presentations was the complementarity of different approaches to agrobiodiversity conservation discussed in the previous section. In various ways, all speakers emphasised that conserving as much biological diversity as possible requires recognising the diversity of conservation practices and strengthening linkages between them. In his talk at the summit, the secretary of the Seed Treaty, Kent Nnadozie, took this argument to a deeper level than previously discussed by highlighting the inevitable reality of *international interdependence* that globalised agriculture has created and cannot ignore, especially in light of the dire social-ecological challenges of the present.

---

9 For an overview of the UN's SDGs, see https://www.un.org/sustainabledevelopment/ (last accessed July 18, 2025).

10 For more information on the SDG Advocates, see https://www.unsdgadvocates.org/about (last accessed July 18, 2025).

In this section, I discuss how the *ex situ* world of conservation frames and grapples with the issue of global interdependence, which has become a subject of increasing concern in recent years. I begin by introducing the notion of interdependence through Nnadozie's talk and then flesh it out based on some of the relevant literature on interdependence in *ex situ* agrobiodiversity conservation. This body of knowledge focuses on what is referred to as international or global interdependence, meaning the interdependence of countries and other national- or transnational-level actors in their access to plant genetic diversity. Based on this understanding, it highlights the need to improve collaboration among globally interdependent worlds.

A small number of rather recent publications that have grown out of a collaboration between *ex situ* conservationists and Indigenous activists translate the recognition of global interdependence in food and agriculture and of the fatal consequences that have arisen from its historical misrecognition into a call for global food system transformation. I go on to unfold this appeal through an extensive narrative expert interview I conducted with one of the Indigenous activists participating in this collaboration, who is also an active member of the first Indigenous community ever to have stored heritage seeds in the Svalbard Global Seed Vault: the Parque de la Papa (English: Potato Park) in the Peruvian Andes. Elaborating on the Potato Park's agri-food system and the cosmology underlying it, I highlight a more-than-human dimension of interdependence, whose recognition and cultivation is crucial for efforts to conserve, cultivate, and share agrobiodiversity among globally interdependent worlds.

To foreground what the human-centred understanding of interdependence in the *ex situ* world of conservation backgrounds, I introduce the notion of *earthly interdependence* as a concept that attends to both international and more-than-human interdependence. Recognising that interdependence is a multiple and situated concept rather than an innocent description of a global reality, I argue that it matters how interdependence is framed insofar as it shapes how global challenges and policies addressing those challenges are defined. On another level, then, this chapter is also a sociological intervention in an ongoing dialogue between interdependent yet fundamentally different worlds. It emphasises the importance of taking other worlds seriously in their cosmological otherness.

The Geopolitics of Global Interdependence

In his talk at the 2020 Svalbard Seed Summit, titled "Global goals and international interdependence: How we all depend on each other to feed ourselves", the secretary of the Seed Treaty, Kent Nnadozie, highlighted some of the ways in which different actors and their efforts to conserve and sustainably use agrobiodiversity essentially depend on one another. The following extract condenses his key arguments and illustrates his understanding of interdependence:

> "These days the greater public [...] thinks that the crops that we have today are just like that, have always been like that. They have lost sight of where it [crop diversity; FV] has come from, that it was never just like that, that it was through human intervention over millennia – deliberate, sometimes trial and error, and sometimes by accident – that we have developed a lot of the crops that we have today in the shape and nature that they are in. [...] They have lost sight and lost connection to the fact that we depend on crops and varieties that come from different parts of the world; and that through changes and interdependence over millennia we have been able to develop them [...]. That is what brings us all together and what we should keep sight of. In practical terms: in the absence of continuous human intervention for the crops we have [...] they will probably die off and that is why we need to continue conserving them as we evolve and develop them, both in *ex situ* conditions as well as *in situ* and on-farm. All countries are interdependent and depend on crops that come from other countries, and no region, no country whatsoever is self-sufficient." (Nnadozie 2020, lecture transcript: l. 12–30)

Two dimensions of interdependence are discernible in this passage, which I will preliminarily characterise as human-centred and more-than-human interdependence. *Human-centred interdependence* refers to the lack of self-sufficiency of human-centred societies or collectives – here nation states and supra-national regions – and their interdependence when it comes to access to a diversity of natural resources. *More-than-human interdependence*, in contrast, signifies the argument that crop plants do not exist or persist as they do and where they do independent of human intervention, as well as its implicit correlate that human life is under threat when crops "die off". Apart from the passages quoted here, in his talk, Nnadozie did not go into the more-than-human dimension of interdependence any further, apparently assuming it as a fact and condition of life.[11] Instead, he focused on the interdependence of human societies, presenting it as a fact as well as an incentive for conservation and agricultural practice and policy more generally. While he used the term "international interdependence", the argument refers to a reality that has emerged through processes of globalisation and their historical forerunners (as will become clear throughout this section). Insofar as these both inhabit and transcend the national, as the sociologist of globalisation Saskia Sassen (2007: 3) has argued, in what follows I use the analytically more accurate notion of "global interdependence" instead (see also Khoury et al. 2015).

In his talk at the summit, Nnadozie argued that humanity's global interdependence is increasingly being recognised against the background of contemporary global threats such as, most acutely at the time, the outbreak of COVID-19.[12] The example of pandemics, he said, illustrates the "common vulnerability" (Nnadozie 2020, lecture transcript: l. 7) that comes with interdependence. In contrast, with regard to crop diversity, interdependence becomes discernible as a chance and "an insurance policy" (Nnadozie 2020, lecture transcript: l. 39) for human life. This choice of words is particularly noteworthy considering the context in which the talk was given. The attribute of providing an insurance policy, in the *ex situ* world, is usually associated with the Svalbard Global Seed Vault (see chapter 3.2). Unlike in the common narrative about the Seed Vault and the collection of duplicates of other genebank collections that it holds, however, it is not the diversity of crops or plant genetic diversity itself that Nnadozie identified as an insurance policy. Instead, it is the practice of sharing and exchanging seeds, which genebanks and the Seed Vault facilitate:

> "[Agricultural] practice has always been based on exchange between farmers who developed [agricultural; FV] products and new varieties. [...] And that exchange also has a very practical, pragmatic purpose. It is not just because of generosity, because they realised their interdependence, that they have done that. It is also an insurance policy, that they are able to share with others. If they face any difficulties, either by natural disaster or by other means,

11 I come back to this dimension in the second half of this section.

12 At the time of the Svalbard Seed Summit (February 24–26, 2020), the COVID-19 outbreak was still a phenomenon considered mostly to affect people in and international travellers from China. The World Health Organization had already classified it as a public health emergency of international concern on January 30, then changed the global risk assessment from high to very high on February 28, but only officially declared it a global pandemic on March 11, 2020.

> they are able to go back and get what they shared with other farmers. So over time, [...] scientists and the technical people have come to realise that interdependence and have tried to push for global mechanisms and global policy frameworks that would capture this reality of interdependence." (Nnadozie 2020, lecture transcript: l. 35–44)

In this passage, Nnadozie establishes a continuity between the earliest and the present forms of agricultural practice and politics. According to this narrative, contemporary efforts to establish international regulations and practices for the conservation, multilateral exchange, and sustainable use of plant genetic diversity are a modern, technoscientific form of the same kind of exchange and mutual support farmers have engaged in since the dawn of agriculture (see also Fowler 1994). The *ex situ* world of conservation with its genebanks, genebank collections, conservation and safety duplication procedures and standards, as well as its policy frameworks, then, is the global-scale apparatus of a globalised world facilitating the same kind of "shar[ing] with others" that farmers have always practised.

Underlying this argument is a conception of the evolution and recent globalisation of a monolithically conceived human culture, particularly agriculture, as a linear process of development and progress that scales up a certain form of life without transforming it along the way. This idea of "scalability" (Tsing 2019) has at least two problematic implications. For one thing, whereas scalability values and creates uniformity, interdependence arises from diversity (Tsing 2019: 145). For another, scalability obfuscates the transformations that forms of life and other things undergo when they expand "as they take on new materials and relationships" (Tsing 2019: 144). While practices of seed-sharing might be as old as agriculture itself, their nature has changed since the Neolithic Age in a similar way to how natures and the (agri)cultures cultivating them have changed throughout their co-evolution with societies, economies, and politics. These changes, which include the de-cultivation of agricultural and biological diversity in the wake of the Green Revolution (see chapter 1.1), have continuously transformed what Nnadozie calls the "reality of interdependence". So, it is a specific reality of interdependence, situated in a specific historical situation, that today's "global mechanisms" (such as the Svalbard Global Seed Vault) and "global policy frameworks" (such as the Seed Treaty) aim to stabilise and leverage.

Hence, it is crucial to consider the historical situation in which Nnadozie emphatically draws attention to global interdependence (here regarding access to a crop diversity required for global food security and adaptation to impending social-ecological challenges) as an "insurance policy" for the future. The current (geo)historical moment is a time in which the future of human life on earth has become precarious in unprecedented ways. It is a time in which the industrialised, technologised, and globalised societies of the Global North have begun to grapple with "the shock of the Anthropocene" (Bonneuil/Fressoz 2017),[13] that is, the dawning realisation that the form of life they have been building for the past 500 years and longer is the most relevant destructive force behind the planetary-scale ecological devastations of the present, including the immense and growing loss of biodiversity. The modern world's Anthropocenic awakening not only confronts it with unprecedented social-ecological challenges regarding a more sustainable political-economic organisation of the larger part of humanity's life with/in nature. By challenging established onto-epistemological certainties, it also triggers a "cosmological shock" (Viveiros de Castro 2015: 61). To a greater extent than other realities and realisations of interdependence, which feminists and other marginalised critical thinkers have been pointing out for decades, the Anthropocene experience topples the dualist logic of dominant Western thinking that underlies reified distinctions such as that between the human and nature, production and reproduction, freedom and necessity (see Plumwood 1993: 43).[14]

13 The notion of the Anthropocene as a new geological era superseding the Holocene was coined by freshwater biologist Eugene Stoermer in the 1980s and popularised by atmospheric chemist and Nobel Prize laureate Paul Crutzen in the early 2000s (Crutzen/Stoermer 2000; Crutzen 2002). Since then, geologists and other natural scientists have been discussing whether and since when "we" are, in fact, living in the Anthropocene and what that implies (e.g. Zalasiewicz et al. 2008; Lewis/Maslin 2015; Steffen et al. 2015b; Benner et al. 2021). The Anthropocene proposal has also attracted a remarkable amount of interest and critique in the social sciences, humanities, and arts (for an overview see Barla/von Verschuer 2022). A salient strand of critique takes issue with the pivotal notion of the Anthropos insofar as it suggests that the main culprit behind the ecological devastations of the present is an undifferentiated humanity. Critics have challenged this by foregrounding specific historically and geopolitically situated forms of organising societal life in and with nature as the decisive geo-transformative forces of the present, and have proposed countless alternative names for the current geohistorical era, such as the "Capitalocene" (Moore 2015, 2016) and the "Plantationocene" (Mitman/Haraway/Tsing 2019; for an overview see Bonneuil/Fressoz 2017; Chwałczyk 2020).

14 One could therefore argue that the Anthropocene experience adds a fifth to the four "great historical wounds to the primary narcissism of the self-centered human subject,

This logic, according to Val Plumwood, is rooted in a principle of "denied dependency" (1993: 41): while both sides of a dualistically structured relation of domination and subordination achieve their identity through that relation, the logic of dualism secures the identity of the master by essentialising both identities and denying their interdependency. A common modus operandi of denial is what Plumwood calls "backgrounding" (1993: 48–49), which refers to the treatment of the subordinated side of any dualism

> "as providing the background to a dominant, foreground sphere of recognised achievement or causation. [...] What is involved in the backgrounding of nature is the denial of dependence on biospheric processes, and a view of humans as apart, outside of nature, which is treated as a limitless provider without needs of its own. [...] This denial of dependency is a major factor in the perpetuation of the non-sustainable modes of using nature which loom as such a threat to the future of western society." (Plumwood 1993: 21)

As the limits of backgrounding Nature come to the fore in the present time-space ironically called the Anthropocene, the Human becomes increasingly unable to deny the dependencies constituting his very existence (see also Hoppe 2024).[15]

It is in this material-discursive context that I propose to situate the *ex situ* world's surging interest in interdependence.[16] In the last decade, *ex situ* conser-

---

who tries to hold panic at bay by the fantasy of human exceptionalism" (Haraway 2008: 11), as chronicled by Freud and amended by Haraway. To the three wounds Freud describes – the Copernican decentring of the earth in the cosmos, the Darwinian decentring of the human in the ensemble of creation, and Freud's own challenge to the primacy of consciousness and reason in the human self-image – she adds a fourth wound caused by cybernetics' and biotechnology's folding of the technological into the organic (see Haraway 2008: 11–12).

15 I capitalise the Human and Nature, here and in what follows, whenever I refer to their (universalist) figuration as per the logic of dualism. The Human is not the human as in *Homo sapiens* but denotes the historically situated figure of the Human, who has crafted himself as universal yet in the image of the white European bourgeois Man (see e.g. Wynter 2003; Mignolo 2015; Jackson 2020). Nature is a co-product of this "invention of the Human" (Mignolo 2018: 153–176) in colonial modernity, through which an other- or less-than-human sphere of dispensable and exploitable life including dehumanised human lives as well as non-human Nature was created (see also von Verschuer/Hartmann/Barla 2023).

16 This surge of interest in interdependence is not restricted to the issue or world of agrobiodiversity conservation. For example, Smith/Clark/Yusoff (2007) provide a compre-

vationists have been discussing global interdependence in PGRFA in a growing number of publications (e.g. Ramirez-Villegas et al. 2012; Ramirez et al. 2012; Khoury et al. 2015; Khoury et al. 2016). Cary Fowler and Toby Hodgkin systematically introduced and distinguished different levels of global interdependence based on the findings of a study on the global availability of PGRFA in a study published as early as 2004. Building on the FAO's first global assessment of *The State of the World's Plant Genetic Resources for Food and Agriculture* (1997) as well as on a few early sources mentioning forms of interdependence (Kloppenburg/Kleinman 1988; Cooper/Engels/Frison 1994; Flores-Palacios 1998), Fowler and Hodgkin (2004: 10.5–10.6) define interdependence along three axes. They show that when it comes to the conservation of PGRFA, today most countries in the world depend on (a) crops that are not indigenous to these countries; (b) 'improved' varieties usually bred with genetic resources from all over the world; and (c) other countries.

The first level of interdependence refers to the fact that all major crops cultivated around the world today are what Fowler and Hodgkin call "botanical immigrants" (2004: 10.2). Like Nnadozie, they argue that plant seeds have migrated around the world since the Neolithic Age and that global interdependence, therefore, is "the result of a kind of botanical chess game that has been played out over millennia, changing the agricultural and food systems of virtually every land" (2004: 10.5). While they do not specify the players of this game, Pat R. Mooney (1983), who coined the notion of the botanical chess game,[17] as well as a number of historians and science studies scholars have comprehensively discussed the role of European colonialism and imperialism in the large-

---

hensive review of academic and political engagements with the notion of interdependence throughout the past century in order to employ it in a discussion of current and future global ecological, economic, and sociocultural challenges. Hoppe (2024) intriguingly discusses how notions of dependency, independence, and interdependence variously inform classical modernist and more recent relational approaches to social theory as well as critical approaches in feminist and postcolonial theory.

17 In "The Law of the Seed", a long read for the journal *Development Dialogue*, Mooney writes: "The history of the colonial era is, to an amazing degree, a history of plant transfers: South to North, or from Latin America to Asia, to Africa to Latin America. Like living pawns in the imperial chess game, coffee, tea, cocoa, rubber, bananas, sugarcane, cotton, and spices, were shifted back and forth over the surface of the globe. Although few could have anticipated it at the time, the botanical chess game has had long-term implications for the development and underdevelopment of the Third World that simply cannot be exaggerated." (Mooney 1983: 85)

scale, often illicit translocation of botanical resources around the globe since the fifteenth century (see Crosby 1972; Brockway 1988, 2002; and more critically Schiebinger 2004; Schiebinger/Swan 2005; Subramaniam 2024). These authors argue that although botanical migration routes have already existed for thousands of years, the global dissemination of all kinds of nonhuman lifeforms has reached unprecedented levels and given rise to unprecedented ecological transformations around the world in the wake of the enormous increase in global travel and resource exploitation since the colonial era.

In line with this historical politicisation of the global interdependence in PGRFA, in a seminal study of the history and political economy of plant breeding and the emergence of the agricultural biotechnology industry, the political scientist Jack R. Kloppenburg provides "empirical justification for the characterization of the North as a rich but 'gene-poor' recipient of genetic largesse from the poor but 'gene-rich' South" ([1988] 2004: 181; see further 180–184). Without contesting the assertion that "[t]here is no such thing as plant genetic independence for either the regions of the North or the South" ([1988] 2004: 182), Kloppenburg politicises the analysis of global interdependence in PGRFA by highlighting the asymmetrical global economic power relations in which that interdependence is embedded – not only historically but also in the present.[18]

A pertinent example of the effect those power relations have in the present day is the second level of global interdependence Fowler and Hodgkin mention: that most countries currently depend on hybrid varieties developed using plant genetic resources from all over the world. While for the larger part of history, plant genetic resources have flowed unidirectionally from the "gene-rich" regions of the Global South to the "gene-poor" Global North, a reverse flow has evolved since the mid-twentieth century. Kloppenburg elaborates this as follows:

> "The initiation of commercial seed exports from the industrial nations to the Third World introduced a crucial qualitative dimension to the established asymmetry of germplasm flow. Plant genetic resources leave the periphery

---

18 For a comprehensive analysis of the political economy of PGRFA and plant biotechnology, see Kloppenburg (1988, [1988] 2004) and in the German-language context Flitner (1995) and Flitner/Görg/Heins (1998). My study backgrounds this analytical perspective in favour of a focus on the onto-epistemological dimensions of global power relations, which have received much less attention to date.

> as the common – and costless – heritage of mankind, and return as a commodity – private property with exchange-value." (Kloppenburg [1988] 2004: 169)

Although the international policy situation has shifted from supporting the global expansion of the seed industry to protecting national sovereignty over PGRFA and advancing farmers' rights as well as principles of access and benefit sharing (especially after the Seed Treaty entered into force in 2004), the asymmetrical power relations between the seed industry and farmers in the Global South persist to this day (Kloppenburg [1988] 2004: 170–175; 2014; Kastler 2016; Nehring 2022). Any comprehensive account of the global interdependence in PGRFA, Kloppenburg emphasises, "must take into account this 'genetic geography'" (Kloppenburg [1988] 2004: 175; see also Flitner 2003) from which the alleged "common bowl" (Kloppenburg [1988] 2004: 175) of PGRFA has emerged.

The third level of interdependence Fowler and Hodgkin distinguish, interdependence in conservation efforts, is a result of the global dissemination of the world's PGRFA. As I elaborated in the previous chapter, medium- and long-term conservation requires periodic regeneration of the conserved samples, ideally in the environment they were collected from. Insofar as genebank collections do not necessarily comprise local species only (and the larger they are, the less this is the case), genebanks depend on cooperation (see Fowler/Hodgkin 2004: 10.6). The commitment to safety duplication as a means of securitisation further increases interdependence in conservation efforts insofar as it requires cooperation because safety duplicates have to be stored in a different location than the original collection, as per the FAO's *Genebank Standards* (FAO 2014: 57–58). This means that there is more to this question than the way in which "[g]lobal interdependence in plant genetic resources provides a strong rationale for proactively conserving and facilitating access to [plant genetic] diversity worldwide" (Khoury et al. 2015: 4), as the authors of a policy brief issued by the International Center for Tropical Agriculture (CIAT) put it in 2015. Rather, insofar as the cultivation, breeding, and conservation of PGRFA are inextricably related processes, the different levels of global interdependence are interdependent themselves.

The discussion of interdependence in the studies of Fowler and Hodgkin (2004) and Kloppenburg ([1988] 2004) focuses on the interdependence of countries and transnational actors in their access to PGRFA, that is, on what I have called human-centred interdependence. The dimension of more-than-human interdependence remains unaddressed. This issue has, however, been considered in some more recent publications bringing together the expertise of scientists from the *ex situ* world and Indigenous conservationists (Khoury et al. 2019; Argumedo et al. 2020; Argumedo et al. 2021).[19] Although they do not systematically make it a subject of analysis, these authors draw attention to the more-than-human interdependence of natures and cultures in their historical and ongoing formation.

For instance, Argumedo et al. (2021: 2) emphasise that "many different cultures around the world became increasingly interdependent with the plants and animals they interacted with through the biocultural processes of domestication". Throughout human history, this has generated an enormous diversity of crop and livestock species and genetic variations within these species, as well as a diversity of knowledges and practices around their cultivation and use. In its entirety, this diversity is encapsulated in the term "biocultural diversity" (Argumedo et al. 2021; see also Martin/Mincyte/Münster 2012). According to Argumedo et al. (2021), the homogenising tendencies of historical and ongoing urbanisation, industrialisation, and globalisation processes – in agriculture and beyond – have resulted in the loss of a large amount of food-related biocultural diversity. They argue that this biocultural de-diversification is the root of many health-related problems. In addition, globalised and industrialised agriculture is "among the most significant contributors to environmental degradation, climate change, and biodiversity loss globally" (Argumedo et al. 2021: 2).

In this biocultural situation, the authors argue, global food system transformation becomes imperative. Recognising and engaging with the diversity and resilience of Indigenous food systems and knowledges, for them, is a crucial component of such a transformation.[20] There are two core reasons for this.

19 The groups of authors behind these papers partially overlap, with three authors involved in all of them: Alejandro Argumedo (Potato Park, Asociación ANDES, and International Network for Mountain Indigenous Peoples (INMIP), Cusco, Peru), Hannes Dempewolf (Crop Trust, Bonn, Germany), and Colin Khoury (CIAT, Cali, Colombia). I interviewed two of these three collaborators for this study: Hannes Dempewolf, whom I have already cited in previous chapters, and Alejandro Argumedo, the interview with whom I discuss below.

20 I follow the self-description of the Indigenous authors in this collaboration as "Indigenous" here. In doing so, I draw on a self-determined understanding of "Indigeneity" that resists "genomic articulations of indigeneity" (TallBear 2013). According to Kyle P. Whyte (2015a, 2015b), a member of the Citizen Potawatomi Nation and a professor at the University of Michigan whose work focuses on climate change, environmental jus-

First, the modern food system depends on access to plant genetic diversity adapted to emerging ecological conditions, whether for a transformative process of re-diversification or for the breeding of new resilient varieties. Most of this diversity, whether conserved *in situ* and on farm or growing in 'wild', i.e. non-cultivated spaces, today exists in what are known as the world's "primary regions of diversity" (Khoury et al. 2016: 2). These are the areas where crops were first domesticated and which have an exceptional richness in domesticated and wild species as well as a high degree of genetic variation within those species. Most of these regions are located in the Global South, often in mountainous or other topographically extreme areas inhabited by Indigenous peoples. Insofar as they are particularly exposed to environmental stressors and ecological challenges and changes, the authors argue that the long-term persistence of Indigenous communities, their self-sufficient food systems, and the species they cultivate and live with in these places prove their resilience and adaptive capacity (see also Khoury et al. 2019: 6).

Herein lies the second reason why Argumedo et al. (2021) consider the sharing of knowledge and expertise with Indigenous peoples to be crucial for sustainable food system transformation in light of developing and challenging

tice, and food sovereignty with a particular interest in Indigenous peoples, "Indigenous peoples" are "the roughly 400 million people in the world who exercise cultural and political self-determination in territories dominated by occupying newcomer nations, such as the U.S. or Ecuador" (2015b: 143). In a keyword entry on "Indigeneity" he elaborates that "the indigenous movement focuses considerably on ecological concerns. Indigenous peoples' political self-determination is often centred on protecting intergenerational systems of placed-based relationships from being obstructed by globalization and other political, social, and economic forces" (2015a: 145). Insofar as they are place-based, Indigenous food systems are as diverse as Indigenous peoples. What unites most Indigenous food systems – and differentiates them from the modern industrialised and globalised food system – is that they are tailored and adapted to the land in which they exist. Accordingly, Indigenous food systems are locally specific ways of producing, distributing, sharing, and consuming food equipped with locally specific modes of adapting to "metascale forces" (2015b: 148) such as climate changes or invasions and disruptions by other (human and more-than-human) populations. While these definitions resonate with the self-conception of Argumedo and the Quechua communities in the Potato Park, it is important to understand them as operative rather than exhaustive definitions. Although broadly articulated, they do not apply to all ways of life identifying as Indigenous. On the diversity of Indigenous ecological relations see, for example, Adamson/Monani (2017) and Bargh (2021).

global ecological changes. The global food system transformation that some of these authors envision by 2050 as spelled out in an earlier paper,

> "centers on safeguarding, creatively using, and celebrating the rich diversity of food plants and animals that sustain all of us. We want to put diversity at the heart of food systems so that they are more sustainable, resilient, and equitable, and so that they are better equipped to provide healthy and delicious food to all." (Khoury et al. 2019: 1)

In the view of Khoury and colleagues, Indigenous food systems illustrate how such a transformation requires the awareness, preservation, and celebration of biocultural diversity, which lies at the heart of their perseverance.

The biocultural model many Indigenous food systems are built on is certainly not entirely compatible with the *ex situ* world's "resourcist cosmovision" (Fenzi/Bonneuil 2016: 78). Rather than considering biological diversity primarily as a resource for the flourishing of human life, many Indigenous cosmologies assume and cultivate a more fundamental relationality of biological and cultural, human and more-than-human life. While this more-than-human dimension of interdependence is mentioned in the collaborative papers cited here, it remains rather marginal. However, it became quite evident in the interview I conducted with Alejandro Argumedo, one of the Indigenous scholars in the above-cited group of authors. Attending to more-than-human interdependence such as spelled out by Argumedo and to its centrality in his community's food system adds a new dimension to the discussion of global food system transformation that I explore in what follows. Awareness of biocultural diversity as well as its preservation and celebration are certainly crucial. But if global food system transformation is to emerge from a meaningful dialogue among diverse worlds it is also crucial to take other worlds seriously in their cosmological otherness. Insofar as Indigenous worlds have a lot to teach other worlds when it comes to surviving worlds' ends (see also Danowski/Viveiros de Castro 2017; Whyte 2017; Yusoff 2018), this is essential not least in light of the existential social-ecological threats the modern world currently faces, to which the global industrial agri-food system has contributed its fair share. I discuss these matters in the following based on the interview I conducted with Argumedo.

The World/ings of the Potato Park

Argumedo is an inhabitant of the Parque de la Papa (English: Potato Park), an Indigenous Biocultural Heritage Area (IBCHA) (Argumedo 2008) covering over 10,000 hectares of the Andes in the region of Pisaq near Cusco in Peru.[21] It was established in 2000 through the combined efforts of six Quechua communities living in the area and with the support of a Cusco-based Indigenous non-profit organisation, the Asociación ANDES, founded among others by Argumedo. The Asociación ANDES developed the IBCHA model along with the Potato Park to describe its "community-led and rights-based approach to conservation which protects and enhances local livelihoods and biocultural diversity using the knowledge, traditions, and philosophies of indigenous peoples related to the holistic and adaptive management of traditional agricultural landscapes" (Argumedo 2008: 45). Today, the Potato Park is home to 6,000 Quechua people as well as over 1,400 native potato varieties and numerous other traditional crops (Asociación ANDES n.d.; Asociación ANDES/Potato Park 2015; Argumedo et al. 2021: 4–5).

In light of the increasing existential threats that global climate change poses to the biocultural heritage conserved in the Potato Park, in 2011 the Potato Park launched a three-year project in cooperation with the Crop Trust to produce and prepare botanical potato seeds for safety storage in the Svalbard Global Seed Vault. In 2015, the Potato Park became the first Indigenous community ever to make a deposit in the Seed Vault (Crop Trust 2011, 2015; see

21 The account of the Potato Park that follows is further substantiated by a thorough engagement with written and filmic documents about the Potato Park, including Argumedo (2008), Argumedo/Yun Loong Wong (2010), Asociación ANDES/Potato Park (2015), and Asociación ANDES/Black Maria (2016). Due to the international travel restrictions issued in response to the COVID-19 pandemic as well as language barriers, I was not able to conduct ethnographic research or further expert interviews. As a consequence, my account of the Potato Park and its world/ings is based on a much less "thick description" (Geertz 2008) than my account of the *ex situ* world of conservation developed in the previous chapter, for instance. Nonetheless, the examination of the Potato Park as a manifestation of a different world, albeit one that is entangled with the *ex situ* world in many ways, offers an insightful contrast insofar as it allows important questions to be raised about how to understand the loss and conservation of agrobiodiversity. In this sense, my discussion of Potato Park with primary reference to the interview with Argumedo is not to be understood as a comprehensive account of his world, but rather as a productive irritation of dominant and all but universal knowledges of and relations to the world.

also Asociación ANDES/Black Maria 2016).[22] Cary Fowler, at that time head of the Crop Trust, commented on the collaboration as follows: "The Potato Park highlights the active role that individual communities play in creating and conserving diversity [...]. This partnership demonstrates the critical importance of the Seed Vault in backing up conservation efforts of all kinds." (Fowler quoted in Crop Trust 2011) In this sense, the collaboration between the Potato Park and the Crop Trust and the Seed Vault is a living example of the complementarity of conservation efforts (see chapter 4.1). Khoury et al. (2019: 5) commend it as "a truly integrated model for conservation" and food system transformation (see also Argumedo et al. 2021: 4). The interview I conducted with Argumedo as well as some of his and his community's publications offer a deeper insight into the world in which the Potato Park's food system and

22 In a storage event following the 2020 Svalbard Seed Summit (see also chapter 5.1), the Cherokee Nation, the largest of three federally recognised Cherokee tribes settled in the northeast of present-day Oklahoma, USA, became the second Indigenous community to make a deposit in the Seed Vault. During the summit as well as in its media coverage, this contribution to the Seed Vault attracted more attention than those of other first-time depositors. What makes contributions by depositors such as the Potato Park and the Cherokee Nation remarkable is not only that they usually pursue a community-based *in situ* approach to conserving their biocultural heritage. Many Indigenous communities are also sceptical of national- and international-level *ex situ* conservation endeavours due to historical and ongoing experiences of dispossession of resources as well as rights and sovereignty over their biocultural heritage. In her study of Native American seed savers and their attitudes towards *ex situ* conservation in general and the Svalbard Global Seed Vault in particular, Sheryl D. Breen points out: "[The] seeds currently being used for commercial food and crop production originated, either directly or indirectly, from the heritage seeds grown and selected by peasant farmers around the world. In accordance with that fact, [...] Native American growers and seed savers stated their concerns that centralized *ex situ* collection of those seeds' genetics would constitute a confiscation of their cultural heritage and denial of their collective rights to use, exchange, and sell their seeds" (Breen 2015: 45–46). The Svalbard Global Seed Vault's unique black box agreement, which guarantees that all deposits, unlike in other genebanks, remain property of the depositor at all times (see chapter 1.5), although not sufficiently convincing in the eyes of all critics (see Breen 2015: 46), has removed some communities' doubts and made collaborations such as with the Cherokee Nation and the Potato Park possible. I focus on the story of the Potato Park here. Marleen Boschen, who investigates seed banking from a cultural studies perspective and who, like me, attended the Seed Summit in Svalbard as a participant observer, intriguingly discusses the ambiguous implications of Indigenous participation in the global seed conservation endeavour following the story of the Cherokee Trail of Tears black bean (see Boschen 2022: 112–153).

Argumedo's vision for food system transformation are situated. In what follows I unfold what I learned from afar from these sources about this world, its practices and principles, as well as its history and cosmology, and raise some questions about predominant narratives in the *ex situ* world of conservation, especially about interdependence and conservation.

I opened the interview with Argumedo by explaining the focus and interest of my research – the Svalbard Global Seed Vault and *ex situ* conservation more generally as a means to meet future social-ecological challenges with regard to food security, as I summarised it at the time. Before I got to asking my first question, he responded with a long reflection on how one frames a topic of research and approaches it based on (possibly implicit) frameworks of thinking, admonishing me to consider, as I would paraphrase it, the constructedness and situatedness of the categories underlying my research.

> "In our tradition, names have very powerful connotations because they connect us with the spirit of the land. When we frame our research – when you frame your research – with names and descriptions or definitions, it's a strong positioning. I think the power of science is the power of naming; and this speaks, in my view, to the proud perception that humans have, which is human exceptionality, that we are the top of everything. We name things, we use them as we name them, we study them as we name them, and we create economies around them because they are our property. I know this is a very old connection but it's a chain that continues and that science continues to turn into a conceptual framework or a worldview – while you may say that it is science that has to be balanced. You have to do your analysis, but from the beginning you are starting from a theoretical belief system that is not science at all; it's a given, it's this divine thing that humans are superior, humans are here to conquer, to create wealth for humans. And so, in the discussions of seeds, we treat seeds as resources, we treat seeds as something totally separated and alienated from people. It is people's rights that we discuss, people's tools that count – based on that naming, because any type of naming is an immediate association of how you do things." (Alejandro Argumedo, Potato Park, 5 November 2020: l. 6–32)

The critical account of Science[23] Argumedo unfolds in this passage resonates with some of the core arguments of feminist and postcolonial science studies, from which I derive the terms of my engagement with his deliberations. Argumedo points out what Haraway has trenchantly described as "situated knowledges" (1988): the idea that knowers, processes of knowledge production, and the knowledge they produce are always and inevitably embodied, emplaced, and engaged.[24] Building on this assumption about knowledge production, he articulates a critique of Scientific representationalism that can be understood in line with a similar argument made by Haraway. As she puts it, scientists give names to the things they are concerned with and then tend to "mistake the names for the things. [...] But the things, in [her] view, do not pre-exist as ever-elusive, but fully pre-packaged, referents for the names." (Haraway 1992: 313) Rather, they become what they are through situated articulations. The power of Science, from this point of view, is the power that lies in taking possession of things by naming them in a way that silences other situated knowledges of their reality (ibid.).[25]

23 I capitalise Science – as I have done earlier with Nature and the Human (see footnote 15 in this section) – to indicate that this argument refers to a historically and culturally specific understanding and practice of science that casts itself as Science proper. Contrary to this assumption, philosophers and historians of science situate the idea of a universal and objective Science in the Western post-Enlightenment world, using more specific terms such as "Western Science" or "Western Enlightenment Science" to highlight its geopolitical and socio-historical origin. Following Indigenous STS scholar Max Liboiron, I prefer to use the term "dominant Science" in order to neither geographically fix Science nor overshadow other Western scientific traditions and reproduce their marginalisation (Liboiron 2021: 20–21 fn 77).

24 Similar arguments have been developed by standpoint theorists such as Sandra Harding (1986, 1991, 2011) and Patricia Hill Collins ([1990] 2000). Haraway's concept of situated knowledges, partly and among other things, is a response to standpoint epistemologies. She recognises a risk in these of romanticising and appropriating subordinated knowledges and hence of both disembodying them and relieving them of the need for critical deconstruction (Haraway 1988: 583–584).

25 Using the example of the colonisation of Australia, Plumwood (2002) develops a critical analysis of "the colonizing politics of place names" (Plumwood 2002: 97), which comes very close to Argumedo's critique of Scientific naming practices. She argues that the convention in Western positivist Science and philosophy of viewing names as "neutral markers without cultural content [...] reflects the concept of the land as neutral, passive and silent; as such, it is an index of the shallowness of relationships to place." (Plumwood 2002: 100) Consequently, she calls for renaming practices "as a

Argumedo's argument that in his world, names, concepts, and definitions "connect us with the spirit of the land" implies that they are situated signifiers of the more-than-nonhuman environment or world in which they were formed and of the relations in which those who use them live with this environment or world. It is important to note here that "land" is a particular category in many Indigenous cosmologies. According to Indigenous science studies scholar Max Liboiron, many Indigenous communities understand land not as landscapes in the universal (even if diverse) sense of "a fixed geographical and physical space that includes earth, rocks, and waterways" (Styres/Zinga 2013: 300–301, quoted by Liboiron 2021: 6 fn 19), but rather as "the unique entity that is the combined living spirit of plants, animals, air, water, humans, histories, and events" (Liboiron 2021: 7 fn 19).[26] Insofar as names, concepts, languages, and stories are formed in the more-than-human context in which they emerge, they are not only indicative of the latter but also shape, as Argumedo emphasises, how one thinks about what one thinks about. Against this background, I understand the sequence with which Argumedo opened the interview as an appeal to avoid uncritically (if unintentionally) reproducing the Scientific "power of naming"; to think critically about the seemingly universal Scientific terminology about seeds and conservation and the human exceptionalist paradigm underlying it, which has made the other-than-human world an appropriable resource for Human purposes by othering parts of the world as nonhuman on the basis of the ontological distinction between the Human and Nature.

Putting it in the words of decolonial philosopher and semiotician Walter D. Mignolo, Argumedo's critique of Science is directed not (primarily) at disputing the content but at "*changing the terms of the conversation*" (Mignolo 2018: 130, emph. in orig., see also 135–152).[27] Instead of talking about the practices,

---

project of cultural convergence, cross-fertilization, reconciliation and decolonization" (Plumwood 2002: 104).

26 This broader definition of what is part of and makes land(scapes) resonates with the notion of the more-than-nonhuman environment I proposed in the previous chapter in order to avoid collapsing the notion of the environment into Nature understood as a sphere ontologically distinct from and surrounding the Human. For an overview of a diversity of Indigenous and non-Indigenous "epistemologies of land", see also Anderl (2024).

27 This critique is decolonial insofar as the terms it challenges constitute the grammar of what Mignolo, following the Peruvian sociologist Aníbal Quijano (2007), calls "the colonial matrix of power" (Mignolo 2018: 111–116, 141–145; see also Mignolo/Segato/Walsh 2024). Synthesised in this concept are the dimensions of an epistemic colonial-

technologies, or politics of plant genetic resource conservation, Argumedo speaks of the technoscientific belief system and terminological framework in which seeds come to matter as plant genetic resources – and in which questions about seeds thus become questions about technology and rights to seeds rather than questions about more-than-human relations.[28] This conversation, according to Argumedo, is a conversation scripted by the "divine", that is, by the allegedly metaphysical and unquestionable rationality of human exceptionalism. It is a conversation that comes from and comes with a world of meaning and power relations, which it activates. However, it is not the only conversation about seeds to be had. The interview with Argumedo shows that there are other worlds in this world, structured by other more-than-human relations and other knowledges, giving rise to other names, ideas, and practices that hold promises of other futures. My aim in tracing these in Argumedo's account of his world is to demonstrate some of the ways in which "[i]t matters what thoughts think thoughts. It matters what knowledges know knowledges. It matters what relations relate relations. It matters what worlds world worlds. It matters what stories tell stories." (Haraway 2016: 35)

Argumedo elaborated on the food system of his community by embedding it in the history of Indigenous life in the Andes and situating this history in the environment in which Andean Indigenous societies developed:

---

ity historically established with European colonialism, which outlasts processes of political decolonisation and persists to this day as "the darker side of Western modernity" (Mignolo 2011). This epistemic coloniality, according to Mignolo, becomes discernible in the benevolent "rhetoric of modernity" (2018: 138–141), which works to secure "the *imperiality of power* [...] not by guns and armies but by the words that justify the use of guns and armies, convincing you that it is for the good, the salvation, and the happiness of humanity" (Mignolo 2018: 140, emph. in orig.) I go into the ways in which the rhetoric of modernity justifies Scientific and more-than-human power relations below.

28 Greta Gaard (2015) develops a similar argument in her queer and posthumanist ecofeminist analysis of the antifeminist tone in dominant responses to climate change. The problem Gaard identifies is a "masculinist bias of traditional western ethics [...] [which] envisions justice-as-distribution of resources among discrete individuals with rights rather than emerging through relationships which shape participant identities and responsibilities" (2015: 20). Instead of "masculinist techno-science approaches" (ibid.) she calls for "queer feminist posthumanist climate justice perspectives at the local, national, and global levels [...] to intervene and transform both our analyses and our solutions to climate change" (Gaard 2015: 20–21). This chapter shares this concern (as does the book as a whole), although the focus here is on colonial onto-epistemological power relations and less on their gendered dimensions.

"In the area where I am, in the Andes, our heritage comes from people that created a very sophisticated civilisation and developed or nurtured a type of agriculture that didn't have influence from the external world. Inca agriculture had such a sophistication in terms of engineering, hydraulics, soil conservation, management of different ecological niches in the verticality – which means managing diversity because crops are adapted to each one of those niches. To do agriculture in this ecology, you had to master diversity, that's the only way you can survive. But in the end, you have to collaborate because alone, wherever you are, you cannot survive because you will be damned to a single crop. But collaborating, you have access to crops that grow in the lower part, in the middle part, or in the higher part. Our system is so defined by altitude that in the higher part you have potatoes and root crops, in the middle part you have grains like quinoas, amaranthus, all types of Andean grains that grow in that part, and in the lower part you have what we call warm crops, like fruits and other types of produce and food that people love because of their sugary and very nutritious content, even though they are not starchy. But what happens with complementarity is that potatoes and root crops have fewer carbohydrates, grains in the middle have proteins, and all the essential minerals and vitamins are in the lower part. So, things move up and down and people have to collaborate to have access to all kinds of food. It is just very natural that in doing so, they have access to a large portfolio of food crops. And knowledge and information also move up and down, pollinators work up and down – adaptation processes to changes in the climate are happening all the time, so experimentation is key, is inherent to the system. Farmers have been experimenting and doing a science of agriculture here for ten thousand years. The Incas were at the last cycle of this process, and they created this very sophisticated civilisation that was rich, the wellbeing was very high, there was no food shortage or any type of elevated poverty that we see now. And they built this empire without money, we didn't know money. So, you can build empires without money if you focus on food. For us, that was the lesson: that we want to test and see how strong this still is in our spirit, so that we can start rebuilding our society – without denying modernity and its connections, because we have been living under colonial rule for five hundred years, but that is a very a small time in our history. [...] If the Incas put food at the centre of their geopolitical ambitions, if the Incas had food at the centre of their science and technology, if the Incas had food at the centre of social organisation, then there are lessons to learn there." (Alejandro Argumedo, Potato Park, 5 November 2020: l. 111–166)

There are at least two important lessons to learn from this passage. The first arises from the argument that Indigenous societies in the Andes and their agri-food systems have been in the making and have adapted to continuously changing environmental conditions for ten thousand years. Based on this, Argumedo argues that Andean agriculture is rooted in a millennia-old experimental "science of agriculture". By explicitly claiming scientific authority for the experience- and experimentation-based biocultural knowledge and practices underlying Indigenous agriculture in the Andes, he challenges the epistemological demarcation between scientific and traditional knowledge through which Indigenous bodies of knowledge have been othered and devalued as non-scientific since colonial times. This positioning critically invokes a long history of epistemic marginalisation that historians of colonial botany (Philip 2004; Schiebinger 2004; Schiebinger/Swan 2005) have shown to be deeply entangled with the material exploitation of Indigenous peoples and lands during the colonial era.

According to Kavita Philip (2004: 171, 181), most historians and political economists tell the story of the colonial appropriation of botanical and other natural/ised resources as a story of economic and political empire-making. She complements this discourse by tracing the connections between the latter and the emergence of the epistemic hegemony of Eurocentric Science. By analysing historical documents such as travel reports, Philip shows that European colonialists appropriated botanical resources deemed economically and medicinally profitable, along with local empirical knowledge about the properties and usages of these plants, on the basis of a narrative of universal humanitarian benefit and progress that concealed their national economic and political interests. This allegedly benevolent appropriation was further justified by relegating Indigenous knowledge to the realm of "unscientific" knowledges on the grounds that it was attuned to locally specific needs and therefore considered "narrow and short-sighted" (Philip 2004: 186). The taxonomy-based systematic science of botany institutionalised in the eighteenth century, in the name of which colonialists prospected for botanical resources, was accredited with observer-independence and universality and thus, it was argued, served the interests of all humanity and nature as a whole (Philip 2004: 185–192; see also Schiebinger 2005). Accordingly, Philip concludes, the demarcation between a supposedly universal Scientific knowledge and 'unscientific' local experiential knowledges is an ideological and rhetorical strategy that serves "to obscure the very specific rootedness of this knowledge in a particular political economy and set of cultural practices" (2004: 192).

Indigenous peoples such as the Andean Quechua communities in the Potato Park, who have "been living under colonial rule for five hundred years" (see Argumedo above) have been affected in different and intersecting ways by the material and epistemic power relations shaping more-than-human life in the area since the colonisation of the Americas.[29] By claiming scientific authority for the millennia-old, traditional, experimentation-based agricultural knowledge the Potato Park's food system is built on, Argumedo contests not merely the fact of epistemic marginalisation but its terms, most importantly a conception of Scientific knowledge that is built on excluding Indigenous knowledges per se. Taking this critique seriously by acknowledging the situatedness of all knowledge does not mean inverting the order of recognition by ascribing a greater degree of validity or objectivity to Indigenous ecological and agricultural knowledges than to dominant Western Science. Nor does it mean uncritically adopting, much less appropriating, Indigenous cosmologies and historical narratives. Instead, it means cultivating awareness of how agri-food systems – in the Andean Potato Park as well as elsewhere – are entangled with the knowledge systems they are embedded in and making this relation a subject of analysis. In doing so, agri-food systems become discernible as material-discursive manifestations of the (more-than-non/human) worlds they have emerged from. To phrase it in decolonial terms with Mignolo (2018: 147), it means "delinking" from the epistemic coloniality of dominant Science and "relinking" with emplaced forms of knowing, being with/in, and cultivating locally specific worlds.

This brings me to the second lesson from the last quoted passage from the interview with Argumedo. His description of how Andean agriculture embraces "the different ecological niches in the verticality" (see above) illustrates how place and ecological conditions matter to the development of food systems and societies. Mountain ecologies such as the Andes are highly diverse ecologies due to the distinct ecological conditions of different altitudes. The diversity found in mountainous regions is not only quantitative, meaning that there is a wide range of different species. Andean crop species are also diverse in nutritious quality, which differs depending on the altitude and respective

29 A number of Indigenous and decolonial movements and thinkers refer to the Americas as Abya Yala, which is a pre-Colombian local name for the American continent (see Walsh 2018: 21). Although it would be an exercise in decolonial naming practices, I do not use this term here because Argumedo and his community do not use it in the sources underlying this chapter.

ecological conditions in which they grow – the diverse "ecological niches" of mountainous regions. What distinguishes the food system that Indigenous peoples have created in the Andes and sustained for thousands of years, according to Argumedo, then, is not the sheer availability of a wide range of biological diversity so much as the capability to "master" that diversity.

Importantly, mastery has a different connotation here than in dominant Western knowledge systems. It does not mean submission and control so much as environmental literacy, (more-than-human) collaboration, and experimentation. Mastering the diversity of mountain ecologies, as Argumedo describes it, means knowing the different ecological niches and cultivating a diverse food system adapted to these niches. It further refers to the more-than-human rather than human-centred and technology-based capability to adapt to ecological changes, which extreme environments experience far more frequently than others. Argumedo further emphasises that for self-sufficient and sustainable food systems and societies to master mountain ecologies in the sense of thriving and persisting in them, collaboration across altitudes is key. Only by sharing food plants and knowledges between the higher and the lower regions does the biodiversity available in the area become accessible to all those subsisting on a diverse food system. Crucially, it is not seeds or 'plant genetic resources' that are shared across ecological niches but produce and knowledge. Insofar as food plants achieve their nutritious quality within the ecological conditions in which they thrive, the diversity of the Indigenous Andean food system Argumedo describes is literally rooted in the Andes' diversity of ecosystems. Sharing diversity, in this context, means maintaining the diversity of ecosystems (as well as the biocultural conditions required to maintain them) and sharing its fruits across different communities.

A concept that has arisen in this biocultural context and that guides food production in the Potato Park offers further insight into the knowledge system and the worldview or world the Indigenous Andean agri-food-system is embedded in: "We work under the umbrella of what we call food neighbourhoods." (Alejandro Argumedo, Potato Park, 5 November 2020: l. 76–77) This concept has found its way into the papers on global food system transformation co-authored by Argumedo and others in the context of the collaborative efforts of Indigenous and *ex situ* conservationists, as discussed above. However, the publications introduce a very vague and seemingly human-centred notion of "food biocultural diversity neighborhoods" as "defined geographic region[s] where community members work together to conserve, use, and celebrate their food-related biocultural diversity" (Argumedo et al. 2021: 4; see also Khoury et al. 2019). In the interview I conducted with Argumedo, he put much more emphasis on the more-than-human relations underlying and encapsulated in the approach. Food neighbourhoods, he argued here, are manifestations of a "holistic" (Alejandro Argumedo, Potato Park, 5 November 2020: l. 91) approach to more-than-human relations that "look[s] beyond the farm into landscapes" (Alejandro Argumedo, Potato Park, 5 November 2020: l. 94). That is, rather than fence in, isolate from other lifeforms, and thus seek to control what is cultivated, the food neighbourhoods approach actively embeds food production in the environment in which farming takes place.

Underlying the effort to produce food in more-than-human neighbourhoods rather than on farms organised around human needs and agency is a cosmology that emphasises more-than-human "interconnection, complementarity, reciprocity" (Alejandro Argumedo, Potato Park, 5 November 2020: l. 91–92):[30]

> "In our understanding [...] there are three ways of learning: one is with the head, which is intellectual; one is with your hands, which is practical; and the other one is with your heart, which is loving, the spirit, the sentiment – and that's the way you connect with a potato or a plant or with an animal more closely, with your relatives, those that live with humans. That is why what we call a human community, in reality, is a community that lives in a neighbourhood. The food neighbourhood concept comes from that because potatoes or other crops are not resources, they are neighbours. We care for each other; agriculture is a practice of caring. We have a Quechua name for it called *uyway*, which means seed but also means something that you nurture; that is literally what *uyway* means: something that you nurture, but that also nurtures you. You care for it, and it cares for you. You have to care for a potato seed because it is going to nourish you with food. It's a mutual, reciprocal relationship." (Alejandro Argumedo, Potato Park, 5 November 2020: l. 680–698)

This passage illustrates that food neighbourhoods, as concept and as agricultural practice, are a manifestation of a cosmology emphasising more-than-human interdependence, in which more-than-human relations are neighbouring or even kinship rather than resource relations. This has very profound implications for the ways in which more-than-human relations are enacted – includ-

30 As elaborated in this book's introduction, I understand and use the notion of cosmology in the anthropological sense (see footnote 11 in the introduction).

ing and especially in food production. Where plants and animals "that live with humans" are neighbours and relatives rather than resources, they are members of more-than-human communities and not parts of a nonhuman environment as distinct from human communities.

The notion of the environment, which remains the most common term for the more-than-human world as conceived in the dominant Western cosmology, presupposes a centre that is its point of reference, which it surrounds and from which it essentially differs.[31] It is a category that acquires meaning within a dualist structure of thinking insofar as the relation of centre and environment corresponds to that of self and other, inside and outside, human and nonhuman, culture and nature.[32] A neighbourhood, in contrast, is not organised around or defined in relation to a centre. It is a community that consists of the neighbours it comprises and the various relations between them. In the Quechua cosmology as Argumedo describes it, the relationship between human and other-than-human beings is not structured through a logic of dualism and hierarchy. Rather, both humans and nonhumans are members of more-than-human communities defined not by resource relations but by affective neighbouring and kinship relations that come to matter in practices of mutual care.

Agriculture, in this cosmological context, is not a practice of cultivating resources on farmland so much as "a practice of caring" for the thriving of food neighbourhoods. Crucially, this care is a more-than-human and mutual form of care, in which all participants, human and nonhuman, engage in modes of caring for one another and, collectively, for the landscapes in which communities or (food) neighbourhoods exist. The Quechua term for seed – "*uyway*, which means [...] something that you nurture, but that also nurtures you" (see above) – intriguingly encapsulates this cosmology.[33] Unlike in the English lan-

---

31 For a similar, more extensive conceptual engagement with the notion of the "environment" and a trenchant critique of its colonial baggage, see More Worlds Collective (2025: 15–28, 57–61).

32 Dualist thinking is characteristic of the dominant cosmology of the West, the Occident, the "heteropatriarchal capitalist modern/colonial world system" (Escobar 2020: xii) or however else one might prefer to specify the world in question, the borders and theorisations of which are subject to both historical specificity and change. For a critical discussion of the various "Great Divide theories", see Viveiros de Castro (2015: 210ff.).

33 Remember: "In our tradition, names have very powerful connotations because they connect us with the spirit of the land." (Alejandro Argumedo, Potato Park, 5 November 2020: l. 6–8; see above)

guage and many other European languages, it signifies not the essence of the vegetal being it describes so much as the more-than-vegetal (or biocultural) relations and practices of care in which this being comes to matter, lives with other (more-than-vegetal) beings, and co-produces continuous relational existence.

Earthly Interdependence and More-Than-Human Mutual Care

The Quechua understanding and practice of agriculture as a form of more-than-human mutual care and the underlying relational ontology Argumedo delineates pertinently illustrate what I introduced above as more-than-human interdependence. Conceiving interdependence in terms of such a more-than-human relational ontology rather than as the international or global interdependence of human-centred communities in their access to plant genetic resources puts a fundamentally different complexion on agriculture and conservation in a globalised world in ecological crisis and the responsibilities that come with this. It raises the question of what worlds and worldings agricultural care cultivates and conservational care sustains.

An instructive source for discussing this question is Maria Puig de la Bellacasa's (2017) engagement with the ethics of care in more-than-human worlds. A premise of her inquiry into more-than-human relations of care, which aligns with the Quechua cosmology as delineated by Argumedo, is that all human and more-than-human life is ontologically and thus unavoidably interdependent (Puig de la Bellacasa 2017: 4). What is crucial for her is that "ecological interdependency is not a moral principle but a lived material constraint" (2017: 160). From this ontological assertion, Puig de la Bellacasa deduces that "for interdependent beings in more than human entanglements, there has to be some form of care going on somewhere in the substrate of their world for living to be possible" (2017: 5). The notion of care expressed in this assumption challenges more common understandings of care as an obligation that arises from an ethical, moral, affective, or even epistemic a priori one can choose to meet or not (2017: 90–91). Instead, Puig de la Bellacasa describes care as a condition of interdependent existence in more-than-human worlds and hence a "nonnormative necessity [of life]" (2017: 70). Moreover, it is "a manifold range of *doings* needed to create, hold together, and sustain life and continue its diverseness" (ibid., emph. in orig.). It is through such doings that specific interdependencies and, along with them, specific, situated, and ever-emergent ethical obligations to care come to matter in the first place. As situated and co-evolving doings, relations of care and interdependence, then, can take many forms.

Following Puig de la Bellacasa and against the background of the differences between world(ing)s of conservation that have become evident in this and the previous chapter, I suggest understanding conservation as a matter of care for more-than-human worlds.[34] This implies that every conservational practice, from *ex situ* to *in situ* and on-farm conservation to the cultivation of sustainable food neighbourhoods, necessarily exerts some form of care. The ways in which the technoscientific practices of *ex situ* conservation (such as preparing seeds for conservation and keeping them viable in *ex situ* conservation environments) are practices of care have been described by a number of scholars working at the intersection of STS and the environmental humanities in recent years (e.g. Wickson 2016; Chacko 2019; Laboissière 2019; Boschen 2022). Argumedo's account of Quechua agriculture as a practice of mutual care in more-than-human neighbourhoods illustrates a different form of conservational care that has co-evolved with a different, more-than-human interdependent cosmology.[35] Taking seriously the idea that *ex situ* conservation and the Potato Park's agri-food system are situated within specific and in many ways fundamentally different material-discursive worlds characterised by different forms and understandings of care means that the essential question is not which one cares 'better' or 'more' (see also Puig de la Bellacasa 2017: 7–9,

34 Puig de la Bellacasa's notion of "matters of care" points to the entanglement of labour/work, affect/affections, and ethics/politics, in the interplay and unsolved tensions of which worlds come to matter (2017: 5). She developed it in a critical discussion of Latour's notion of "matters of concern" (Latour 2004). Latour self-critically proposes the latter to expand his earlier analyses of the practices (labour/work) driving the social construction of "matters of fact" by the concerns (affect/affections) motivating these practices. Next to the idea that a thing can be both real and constructed, Latour highlights that things are also lively themselves insofar they gather and hold together collectives, for whom they are a common concern (see also chapter 3.2). He thus emphasises that things become what they are in and through networks of agencies and concerns. While Puig de la Bellacasa (2017: 30–48) welcomes Latour's effort to acknowledge and symmetrise distributed agencies and affects in (knowledge) politics, she criticises his reluctance to embrace the questions of power, marginalised and erased voices, and radical politicisations often raised by this kind of politics, which she aims to rectify with her notion of "matters of care".

35 For an account of more-than-human care as a mutual relationship enacted by both humans and other-than-humans, see also Silberzahn (2024).

42–43). Insofar as answering this question would require predetermining universal standards about what good or enough care is, it would disentangle practices of care from their respective worlds. Instead, thinking with Puig de la Bellacasa, the more relevant questions to ask about different forms of conservational care are: "'For whom?' but also 'Who cares?' 'What for?' 'Why do "we" care?' and mostly '*How* to care?'" (Puig de la Bellacasa 2017: 61, referencing Star 1995, Bowker/Star 1999, Bowker/Star 2007, emph. in orig.).

Addressing these questions brings into view the more-than-human relations that different forms of conservational care cultivate. According to Puig de la Bellacasa, this is vital (in the literal sense of the term) because how exactly relations of care are enacted has profound, material consequences for the world that comes to matter in these relational practices:

> "That [care] requires doing something indicates [...] that a more than human world's degree of livability [...] might well depend on the caring it manages to realize. Standing by the vital necessity of care means standing for sustainable and flourishing relations, not merely survivalist or instrumental ones." (Puig de la Bellacasa 2017: 70, emph. in orig.)

In this sense, it matters – materially as well as in that it is significant – what kind of more-than-human relations agri-food systems (including conservation infrastructures) are built on, because the world they cultivate is the world they sustain. Especially in light of the continuously escalating social-ecological devastations of the present, "the question of what worlds will (our) care become enrolled in sustaining becomes even more acute" (Puig de la Bellacasa 2017: 205).

As the previous chapter showed, in the *ex situ* world of conservation technoscientifically trained conservationists carefully conserve the largest possible diversity of PGRFA so as to ensure, in a combined effort, that farmers and breeders around the world have access to as many genetic resources as possible so that they can adapt crops and agricultural production to changing environmental conditions. These efforts primarily anchor the promise of future food security in, and ultimately work to preserve or extend the present of, a technoscientific agri-culture organised around principles such as technologically facilitated adaptation, genetic optimisation, and economic profitability. Insofar as *ex situ* agrobiodiversity conservation leaves the causes of changing environmental conditions largely unaddressed, it fails to account for the fact that they are rooted, to a large degree, in this same agri-culture and its resource-extractivist relation to the nonhuman world. Put differently, by reiterating the failure of modern agriculture to recognise more-than-human interdependence, *ex situ* conservation conserves instrumental, survivalist relations to the nonhuman world rather than cultivating more-than-human relations that might sustain or even increase earthly liveability. In the Potato Park, in contrast, farmers who derive their knowledge and proficiencies from an experimentation-based ancestral tradition of place-based agriculture strive to care for more-than-human food neighbourhoods together with their vegetal and animal relatives in order to preserve mutually life-sustaining relations.

Having said that, the degree of liveability the local communities manage to realise in the Potato Park, howsoever sustainable and flourishing the more-than-human relations they cultivate, also hinges on ecological and climatic conditions beyond local control. In times of global climatic and geo-ecological change, what is now quite commonly referred to as the Anthropocene, the ecological conditions of life are changing all over the world. The socio-economic causes of the "great acceleration" (Steffen et al. 2015a) of earth system changes since the 1950s, which many consider to mark the beginning of the Anthropocene (see Lewis/Maslin 2015), may originate in the centres of industrial capitalism. Nonetheless, in that they transform climatic and ecological conditions of the whole earth system, they affect all life on earth. Accordingly, although the Potato Park's agri-food system is designed to be self-sufficient, it is inextricably entangled in a planetary-scale web of what I propose to call *earthly interdependence*.

The notion of earthly interdependence is a critical intervention into the debate about *global interdependence* discussed at the beginning of this chapter. Without contesting the latter, the notion of earthly interdependence shifts the focus from the *globe* to the *earth* as the key figure of connection. This shift is based on the premise that "the globe" and "the earth" are more than innocent, distinct designations of a given entity or reality preceding its designation. Following Bruno Latour and Donna Haraway, I understand them as "figures" (Latour et al. 2016: 355) that in/form the ways we relate to and inhabit or *world* the world (see Haraway 2016; Latour 2018: 41 referencing Haraway); "figurations" (Haraway 1997: 11) that perform what they presume and "can be inhabited" (ibid.) themselves as a way of meddling in the worlding of the world.

The globe is the anthropocentric and human exceptionalist horizon of modernisation and globalisation. Latour argues that after centuries of being the guiding figure of modern world-making, in light of the rapid reaching and crossing of planetary boundaries it has become evident that "there is no Earth corresponding to the infinite horizon of the Global" (Latour 2018: 42; see also Latour et al. 2016; Latour 2017: 111–145). In other words, there are earthly limits to the modernist ideas of progress and development that can no longer be ignored (Latour 2018: 18). The disappearance of the globe as a horizon of world-making, for Latour, is a loss that opens up new horizons. Whereas others (re)turn to the notion of the land, he argues that there is no going back because modernisation and globalisation have irrevocably transformed the reality of the land. Both the Local and the Global are "utopias, in the etymological sense of the word, places with no *topos*, without earth and without land" (Latour 2018: 41).

Instead, Latour appeals to Humanity to become "earthbound" (2017: 248–253). Becoming earthbound, for him, means becoming attached to Earth in a way that fundamentally differs "from the way humans were attached to Nature" (Latour 2017: 281). It means ceasing to treat the earth "as an inert and mute object [n]or as supreme judge and final arbiter" (ibid.); moving beyond a dualistic understanding of Humanity and Nature, according to which Humanity is determined to either exploit Nature's resources or fall prey to its forces – to either dominate or be dominated by Nature. Instead, "the Earthbound understand that, contrary to what the Humans have never stopped imagining, [...] *they are not alone in the command post*" (Latour 2017: 283, emph. in orig.). The notion of earthly interdependence and the conceptual shift away from global interdependence that I propose is a response to this call to account for the fatal premises and repercussions of the modern world-making endeavour by recognising earthly boundaries and earthboundedness.

Another important influence for both Latour's and my turn to the earth is Haraway's thinking, as articulated above all in *Staying with the Trouble: Making Kin in the Chthulucene* (2016). In this book, Haraway embraces the figure of the earth so as to meddle constructively and responsibly (or rather response-ably) in its figuration.[36] Intervening in the scholarly debate on the Anthropocene,

---

36 Haraway's thinking and writing are characterised by an elective affinity with and at the same time a critical distance to Latour's work. She appreciatively describes this relation in a eulogy for Latour published in a special issue of *Social Studies of Science* that assembles a series of obituaries by his academic companions. One of the main disagreements between Haraway and Latour, which she identifies here alongside a number of shared influences and strivings, is that in thinking about technoscientific fact- and world-making, "[he] liked trials of strength and insisted on writing and inscription as the proper names for what was happening. Those moves have been extremely productive for gen-

she proclaims the "Chthulucene", which denotes the beginning age of the "chthonic ones", the "beings of the earth" (2016: 2). Unlike the Anthropocene and most of the countless neologisms critically rearticulating it in an effort to replace the eponymous Anthropos and its homogenising tendencies with a signifier more accurately encapsulating the main culprit for the current predicament, the Chthulucene does not circumscribe an era or world that is responsible for the social-ecological devastations of the present. Instead, it imaginatively prefigures "a kind of timeplace for learning to stay with the trouble of living and dying in response-ability on a damaged earth" (ibid.). The Chthulucene is a creative reminder "that the established disorder is not necessary; another world is not only urgently needed, it is possible, but not if we are ensorcelled in despair, cynicism, or optimism, and the belief/disbelief discourse of Progress" (Haraway 2016: 51).

Staying with the trouble, for Haraway, means "acknowledging the extent and seriousness of the troubles [we are in]" (2016: 4) while at the same time resisting the all-too-common sentiments of apocalypticism and game-over cynicism, on the one hand, and fantasies of techno-salvation, on the other hand. Insofar as totalising narratives such as that of the Anthropocene and many of its rearticulations effectively encourage such responses, she emphatically argues that "we *must* change the story; the story *must* change" (2016: 40, emph. in orig.). That is, in order to become capable of developing meaningful responses to "the horrors of the Anthropocene" (Haraway 2016: 3) – to become capable of response, able to respond, response-able – we need stories, narratives, figures, terms and suchlike that help us realise that "human beings are with and of the earth" (2016: 55), that "[w]e become-with each other or not at all" (2016: 4). For Haraway, changing the story – or the terms of the conversation, as decolonial thinkers say (see above) – "in times that remain at stake, in precarious times, in which the world is not finished and the sky has not fallen – yet" (2016: 55), means changing not only how we know, but how we live in, relate to, and become with the world – how we *world* the world. In this sense, responding to the troubles of the present means both rethinking and re(con)figuring relations with the world – changing the stories we tell about more-than-human relations and making kin beyond species boundaries.

---

erations of STS scholars. But, with many other feminists in STS, I always wanted something other than trials of strength and tropes so close both to war and to one-eyed monsters, something more like caring and diverse collective composing and decomposing, something with a sensorium not restricted to vision or writing." (Haraway 2023: 2)

My proposition to think about global interdependence as earthly interdependence builds on Latour's call to become earthbound and on Haraway's call to stay with the trouble of "living and dying well with each other in the tissues of an earth whose very habitability is threatened" (2016: 132). Without disputing the global interdependencies that centuries of modernisation and globalisation have created, my aim is to foreground – in addition to these – the more-than-human entanglements of human life that the cosmology of the modern world has consistently "backgrounded" (Plumwood 1993: 48–49). I describe these entanglements as "earthly" because they refer to the locally specific earthboundedness of all beings and ways of life, as well as to the interdependence of all earthbound life through its earthboundedness. Insofar as recognising earthly interdependence means changing the story or the terms of multispecies life on earth, it requires delinking from the *ex situ* world's "resourcist cosmovision" (Fenzi/Bonneuil 2016: 78) and the extractivist practices this informs. This does not mean renouncing *ex situ* conservation altogether so much as recognising *ex situ* conservation and the global safety net for plant genetic diversity that the Svalbard Global Seed Vault offers as valuable emergency measures in the face of tremendous (agro)biodiversity loss without mistaking them for a sufficient solution.

This understanding of the Seed Vault also finds expression in how Argumedo describes its value for the Potato Park:

> "Svalbard could be our life insurance. The way the world is going – and I think this pandemic just amplifies all our problems – we know that climate change is not going to stop. [...] Maybe we are past the tipping point, but maybe Svalbard, at some point, is going to give us back our potatoes because maybe we will survive all the catastrophic situations and then we can rebuild from the Svalbard Global Seed Vault." (Alejandro Argumedo, Potato Park, 5 November 2020: l. 506–516)

The interview passage illustrates that Argumedo and his community recognise the Seed Vault's potential value for saving the biocultural heritage of the Potato Park (and that of other worlds in this world), while remaining aware of the fact that the Seed Vault is a subjunctive space and that the futurity it promises hinges on overcoming the "catastrophic situations" unfolding in the present. Against this background, the Seed Vault did not assume a key role in what Argumedo talked about in the interview I conducted with him, and the story of the Potato Park's collaboration with the *ex situ* world does not end there.

Without contesting the importance of the Seed Vault as a safety net against the loss of plant genetic diversity in and for a globally interdependent world, the story of the Potato Park shows that the challenges of an interdependent world go beyond global interdependence in access to PGRFA. It illustrates the ontological and vulnerable condition of earthly interdependence, that is to say the more-than-human interdependence of all earthly life in local spaces as well as on a planetary scale, as well as the fact that for earthly interdependence to be "sustainable and flourishing" (Puig de la Bellacasa 2017: 70; see above), sustainable and flourishing more-than-human relations need to be cultivated. Insofar as agri-food systems assume a critical role in this, global agri-food system transformation is a crucial element of Argumedo's vision for the future articulated in the collaborative papers cited above as well as the interview. The principle of more-than-human mutual care guiding food production in the Potato Park is an instructive and generative principle for such global agri-food system transformation. As Argumedo introduced it, it is a place-specific, field- and time-tested way of cultivating non-instrumental more-than-human neighbouring or kinship relations.

Crucially, neither Argumedo's nor my intention here is to scale this particular, situated form of organising more-than-human life up to the global level and turn it into a promise for a new, sustainable global agri-food system.[37] As I have previously argued with Tsing (2019), the idea of scalability assumes and enacts decontextualisation rather than place-specific situatedness, which is a crucial part of the Potato Park's agri-food-system and cosmology. In this sense, the 2050 Vision for Global Food System Transformation (Khoury et al. 2019) that Argumedo and his Indigenous and *ex situ* world colleagues co-articulate envisions "a global village of 'diverse food neighbourhoods' spread throughout the planet, comprised of local custodians in regions with longstanding food diversity traditions, productively interconnected with one another and with the global food system as a whole" (Khoury et al. 2019: 4). While this vision does indeed scale the food neighbourhood concept up to a global level, as Argumedo set it out in the interview this does not mean modelling a singular global food system after the Quechua food system. Rather, the global village or neighbourhood of diverse food neighbourhoods he imagines arises from a transforma-

37 Accordingly, I will not spell out what more-than-human mutual care looks like in practice (nor can I, as my empirical material does not allow me to do so). Rather, I conceive it as a specific, situated modus operandi of an ontology of earthly interdependence that has enabled me to spell out the latter concept.

tion of global food systems (in the plural) based not on one (whichever) model but (again in the plural) on

> "mutually beneficial cross-fertilisation of ideas and experiences, and respect that those traditions have the right to continue evolving in their spaces. [...] People have been practising agriculture in the world for more than ten thousand years and in places they have the knowledge about, they name things in their own language, they have created very effective and sophisticated systems. So, let them flourish, let them create solutions wherever they are and like a quilt start weaving, create this beautiful colourful tapestry where everything is interconnected but also maintains its own identity." (Alejandro Argumedo, Potato Park, 5 November 2020: l. 785–787; 798–807)

This vision for global food system transformation draws strength from situated earthboundedness and cosmological pluralism rather than from universalist decontextualisation and the reification of ontologies and epistemologies. A global food system conceived and cultivated as a neighbourhood of diverse agri-food systems situated in diverse more-than-human worlds is a vision for global food system transformation that takes the principle of more-than-human mutual care to the level of both earthly and global cooperation. Agro-bio(cultural)diversity conservation, then, means more than extending the present of a specific world on the brink of extinction, as I have shown to be (at least part of) what the *ex situ* world of conservation strives for. It means assembling and conserving not only a diversity of biological resources but a diversity of more-than-human worlds and modes of worlding.

5. Conserving (Hi)Stories, Cultivating Futures

> It matters what matters we use to think other matters with; it matters what stories we tell to tell other stories with; it matters what knots knot knots, what thoughts think thoughts, what descriptions describe descriptions, what ties tie ties. It matters what stories make worlds, what worlds make stories. (Haraway 2016: 12)

Seeds do not exist irrespective of the ecological and more-than-nonhuman relations in which they come to matter, that is, in which they assume both significance and materiality. It is in this sense that the diversity of seeds conserved inside the Svalbard Global Seed Vault comes with a diversity of ways of life – or of worlds and worldings. Throughout this book I have argued that by conserving seeds, the Seed Vault assembles the worlds and worldings that come with these seeds, thus creating an arena in which what the Seed Vault is and does is negotiated. In the previous chapter I contrasted the *ex situ* world of conservation, which the Seed Vault has emerged from and serves as a safety net for, with the world that the Potato Park in the Peruvian Andes, the first-ever Indigenous depositor to the Seed Vault, is situated in. I have shown that while these worlds meet and overlap in certain practices as well as in the collaborative spaces they have created, they also fundamentally differ with regard to the world(ing)s they enact and seek to conserve. Whereas the *ex situ* world knows, uses, and conserves seeds as extractible units of "life itself" and as "plant genetic resources" that farmers, breeders, and scientists use to cultivate food and biocapital, in the Potato Park humans know and live with plants and seeds as relatives and neighbours in an agri-food system built on relations of mutual care.

Despite these fundamental differences and partial incommensurability, both worlds know that they are also interdependent. In the *ex situ* world of conservation, *global interdependence* is considered an asset insofar as global collaboration promises access to a broad range of agrobiodiversity. Discussing the issue of global interdependence through the Potato Park and understanding the latter as a specific mode of knowing, being in, and enacting the world rather than merely an environment harbouring a great plant genetic diversity, I have shown that interdependence is not merely a question of global access to PGRFA but an existential vulnerability and responsibility arising from a globally shared earthboundedness of all forms of human and more-than-human life. I have therefore proposed the notion of *earthly interdependence* as a way of grasping interdependencies both across and with the earth. The Potato Park has served as an instructive example of what agrobiodiversity conservation under conditions of earthly interdependence can look like. For one thing, it is a world and mode of worlding in which humans and other-than-humans live in relations of mutual care. It is also committed to cultivating mutually beneficial relations of cross-fertilisation with other worlds and modes of worlding it interdepends with, such as the *ex situ* world of conservation. The juxtaposition of these interdependent worlds and modes of worlding has served as a pertinent example of how it matters – to phrase it in a Harawayian manner – through what ideas, experiences, knowledges, cosmologies, and worlds issues such as agrobiodiversity loss and conservation are understood, addressed, discussed, and tackled.

The present chapter turns to some of the stories, narratives, and imaginaries the Seed Vault assembles. The underlying assumption is that worlds are not only (porously bounded and fluctuating) assemblages of actors and the concerns, practices, and technologies they share. They are also what Situational Analysis calls "universes of discourse" (Clarke/Friese/Washburn 2018: 18, referencing Strauss 1978) or, to put it in terms of decolonial theory, "universes of meaning" (Mignolo 2018: 196) that come with specific, historically situated sets of stories, narratives, and imaginaries. These are indicative of the relations that make those worlds and at the same time means of ongoing world- and (hi)story-making themselves. Attending to stories, narratives, and imaginaries, therefore, is a way of attending to how specific relations come to matter, to how worlds are made, and to the figurative or political potential of conceiving worlds and relations as achievements.

The first section turns to a persistent set of stories and imaginaries about the Svalbard Global Seed Vault as a "doomsday vault" or a "Noah's Ark" for the world's seeds. I discuss how these biblical and apocalyptic imaginaries co-emerge with technoscientific promises of salvation associated with the Seed Vault, which the people behind it both seek to dispel and co-produce. In the second section, I contrast the *ex situ* conservationist promise of undoing agrobiodiversity loss with a different set of stories and imaginaries assembled by an art project drawn to Svalbard by the Seed Vault. Through an analysis of this art project as a performative challenge to what *ex situ* conservation environments conserve and what future world(s) they serve, I develop a reading of the art project's experimental form of conservation as a commemorative means of mourning and learning to live with ecological loss.

5.1 "And No, It's Not About the End of the World!" Apocalyptic Imaginaries and Technoscientific Promises of Salvation

> "'Doomsday Vault' Opens" (Nature 2008)
> "Preparing for Doomsday" (Scientific American 2008)
> "The (End of the) World Seed Vault Opens" (Wired 2008)
> "Barroso to Open 'Noah's Ark' for Seeds" (EU Observer 2008)
> "The Safe of the Apocalypse: Noah's Ark where all the Seeds of the World are Saved" (Le Monde 2010)

Newspapers around the world reported on the opening of the Svalbard Global Seed Vault in February 2008 with headlines such as the above. In doing so, they contributed to an emerging salvationist narrative that continues to dominate the international media coverage of the Seed Vault to this day. Next to media reporting, according to the environmental anthropologist Can Dalyan (2021), this narrative goes back to words uttered by the honorary guests at the Seed Vault's opening ceremony. José Manuel Barroso, then President of the European Commission, solemnly declared the Seed Vault a "frozen Garden of Eden" (Dalyan 2021: 184); Jens Stoltenberg, at the time Prime Minister of Norway, referred to it as "the Noah's Ark for securing biological diversity" (ibid.). Referencing the first book of the Christian Old Testament or the Hebrew Bible, these metaphors evoke an imaginary of the seed diversity conserved in Svalbard as God's creation and of the people behind the Seed Vault as its devout custodians. This biblical and salvationist imaginary also informs the more popular and widespread story of the doomsday vault, which alludes to the story of the Apocalypse in the last book of the New Testament in the Christian Bible. The doomsday narrative suggests that the Seed Vault guards a salvific collection of seeds securely preserved in the Arctic permafrost until after any possible apocalyptic event in the future, when they will serve to restore agricultural production and food security.

Ever since the Seed Vault was opened on February 26, 2008, the partners behind it have unceasingly been striving to counter this image by underlining the Seed Vault's pragmatic purpose of securing ongoing conservation efforts in the present. In a short paper published as early as March 2008, Cary Fowler, the founder of the Seed Vault, commented: "Popular press reports about the 'Doomsday Vault' […] typically mask the complexity of the endeavor and, if anything, underestimate its practical utility." (2008a: 190) Nevertheless, countless press reports and various other sources of knowledge about the Seed Vault published over the course of the years that followed, including by and in cooperation with the partners behind the Seed Vault, upheld the doomsday narrative.[1]

The global media coverage of the international seed summit and seed storing ceremony held in Svalbard on the occasion of the Seed Vault's twelfth anniversary in February 2020 (see chapter 4.2) used the biblical metaphors in approximately fifty per cent of the headlines.[2] This is surprising, insofar as the partners behind the Seed Vault repeatedly attempted to dispel the doomsday narrative throughout the summit. During a press briefing on facts and figures about the Seed Vault on the opening night of the three-day event, Hannes Dempewolf from the Crop Trust accentuated its very mundane and practical role within the *ex situ* world of conservation. He emphatically repeated the appeal

1 Despite counterclaims such as the one by Fowler quoted here, the partners behind the Seed Vault also played their part in reproducing the doomsday narrative in the early years after its opening. A possible reason for this is the international attention needed to attract further financial resources and seed deposits, which the narrative certainly helped raise. This narrative was also discernible in public relations activities of the Crop Trust such as the tagging of early pictures of Seed Vault posted on *flickr* (an online community for sharing amateur and professional high-resolution photos and videos) with "doomsday vault" as a caption – an observation I owe to Marleen Boschen.

2 According to a media coverage overview provided by the organisers to all attendees after the summit, out of 95 news pieces published in the international press by March 2, 2020 (two of which are unintelligible to me language-wise), 39 refer to the Seed Vault as a "doomsday vault", six as a "Noah's Ark", eight characterise it as preparing for an "apocalyptic" scenario, whereas only five present it as responding to "climate change fears" or an impending global catastrophe or disaster, and two as a measure against "extinction".

in his talk at the summit the next day, this time undergirded by a presentation slide reading: "And no, it's not about the end of the world!" (Dempewolf 2020: 12) Instead, he said, the Seed Vault serves the "many, many genebanks around the world that need attention and support" and that "[i]n the end it's all about food and those who grow it" (Dempewolf 2020: 13).[3] Throughout the summit, other speakers reiterated the message that the Seed Vault ultimately serves to sustain food production in the here and now, focusing attention on the challenges associated with this task. Nevertheless, these efforts were not able to dispel the popular story of the Seed Vault as a reservoir of resources for (post)apocalyptic times.

While the excessive embrace of biblical metaphors in reports on the Seed Vault can certainly be explained at least in part through economically motivated strategies of the global media industry (such as the clickbait logic of the internet era), it is not restricted to media reports and therefore not reducible to media logics. A significant number of the social scientists and cultural studies scholars who have written about the Seed Vault uncritically reproduce the doomsday narrative in their accounts (Qvenild 2008; Fava 2013; Gan 2015; Pellegrini/Balatti 2016; Nadim 2018; Grigoli 2024). Others mention the biblical metaphors the Seed Vault is often narrated through without adopting them, yet do not subject them to critical scrutiny either (van Dooren 2009b; Roosth 2016; Heatherington/Perley 2017; Kroløkke 2019; Dalyan 2021). To this day, to my knowledge, there exists no comprehensive analysis of the Seed Vault's biblical storying and the implications of these popular narratives. Following Donna Haraway's argument that "it matters what stories we tell to tell other stories with" (2016: 12), this chapter aims to address this gap and scrutinise the ways in which the biblical stories around the Svalbard Global Seed Vault matter, that is, become not only means of accessing worlds but also modes of world-making.

Foregrounding the doomsday narrative in particular, which is by far the most frequently adduced metaphor, I bring into view how the partners behind the Seed Vault reject it in their narrative accounts of the Seed Vault but nevertheless performatively embraced and invoked religious storytelling in the religiously-symbolically saturated mise-en-scène of the Seed Vault during a seed storing ceremony on the occasion of its twelfth anniversary. I begin by developing an interpretation of this event, which I attended as a participant observer,

3 For a discussion of the implications of such arguments, see chapter 4.

through Roy Rappaport's (1979, 1999) sociological theory of the performativity of rituals. I show that the ritualised character of the event enacted a kind of collective commitment to the cause of agrobiodiversity conservation that a scientific context such as the preceding summit cannot evoke in the same way. I argue that the event contributed to the persistence of these stories insofar as the ceremony not only symbolically invoked religious interpretations but also enacted a relation to the nonhuman world – here represented by the seeds deposited in the Seed Vault – that deeply resonates with the biblical stories about the Seed Vault. Through a closer engagement with the biblical story of the Apocalypse, I show that the rejection of the "doomsday vault" narrative is based on a simplistic, literalist understanding of the trope of the apocalypse that universalises the world assumed to come to an end and depoliticises that end. This conception of the rejected apocalypse, I conclude, is connected to a techno-optimistic salvationism infusing the *ex situ* conservationist endeavour of extending the present of the *ex situ* mode of world-making. In this sense, the Seed Vault is very much "about the end of the world", albeit the looming end of a particular world.

The Religious Performativity of a Seed Storing Ritual

The day of the 2020 Svalbard Seed Summit, which I have addressed in the previous chapter, culminated in an anniversary ceremony in front of the Seed Vault. It mainly revolved around the ceremonial storing of the largest quantity of seeds since 2008. 35 genebanks, including seven first-time depositors, brought 188 boxes containing 65,119 new seed samples into the Seed Vault, thus increasing the total number of accessions inside the vault to 1,050,000 (see Crop Trust 2020). After one and a half days in the familiar setting of a scientific conference with talks, slide shows, Q&A sessions, and networking in coffee breaks, and especially after the repeated efforts to disambiguate the conception of the Seed Vault as a pragmatic conservation facility, I was struck by the ritualised ceremonial character and religious aesthetic of this event. The following vignette illustrates my experience of the seed storing ceremony:

The ceremony begins at 5 p.m. The temperature is -20°C. The dress code is snowsuit.

In the darkness of the Arctic night, the artwork embellishing the top of the Seed Vault's concrete entrance glows and glistens in a turquoise light. The snowy hillside behind the entrance building is bathed in blue spotlights, the area in front of it divided and illuminated by ice cube-shaped lanterns forming a semi-circular line on the snowy ground assigning spectators their place. When I enter the crowded scene, one of the ice cubes is already broken, presumably by an inattentive snow boot.

The ceremony is opened by an all-male choir dressed in sailors' uniforms marching up to the site singing and forming a line on the right of the Seed Vault's entrance; on the left, the SDG advocates, as honorary guests, overlook the ceremony from a little podium that is covered in dead animals' skins. A van backs up to the crowd slowly coming to a halt at the top of a line formed by the genebank representatives who are here to make a deposit.

When the doors to the Seed Vault open, I hear a humming sound, and it takes me a moment to realise that this is not a technical buzz coming from inside the vault but a hum coming from the choir. It will last throughout the seemingly endless row of one genebank representative after another taking their box out of the van, carrying it to the entrance of the vault, handing it over to NordGen staff who carry it inside the vault, being loudly announced and solemnly handed a certificate by the Norwegian Minister of Food and Agriculture, accompanied by thickly gloved muffled applause and a flurry of camera flashes, all under the eyes of the honorary guests and other spectators who appear increasingly uncomfortable in the freezing cold unrelentingly creeping under no matter how many layers of warm clothing.

Thirty-five ceremonial deposit-handovers later, the Seed Vault's doors close and the freezing crowd quickly disperses. The way back to the buses that have brought us here from the summit venue and will now take us back into town leads through a guard of honour formed by the still-singing sailor choir. I cannot help but sweep a curtsy at the end. While the childish and frisky energy that arises within me towards the end of the ceremony might also just be an effect of the staggering cold, I am much more inclined to attribute it to the mise-en-scène of the event. Overcome by childhood memories of sitting through seemingly endless church services in unheated cathedrals where I can see my breath condense before my eyes as I sing, choking siblingly laughter under the severe eyes of older family members and other churchgoers – memories that evolve, as soon as the ceremony is over, into the disenchanted sense of surrealism my childhood recusancy equally evolved into – I cannot help but feel that what I experienced here is not a pragmatic technoscientific conservation procedure so much as a religious service.

Lights, choral singing, ritualised offerings – the dramaturgy of the event temporarily seemed to turn the illuminated Seed Vault entrance into an altar and the storing of boxes of seeds into a ritual reminiscent of Christian rites such as Holy Communion or Thanksgiving.[4] Against the backdrop of the efforts throughout the summit to dispel the popular religious narratives the story of the Seed Vault is often cloaked in, two questions arise that will guide my discussion of the meaning of the event in what follows. First, I ask what the symbolically saturated rituality of the seed storing ceremony means, and does, in the technoscientific context of the seed summit and agrobiodiversity conservation more generally. Second, venturing the hypothesis that the ritualised and symbolically charged seed storing ceremonies occasionally conducted on anniversary events such as the one described here performatively co-produce the salvationist stories of a doomsday vault or a Noah's Ark for seeds, I raise the question of how it matters that the technoscientific story of the Seed Vault is told, as it is, with the help of biblical stories.

The study of rituals has a long tradition in the sociological analysis of social action. The most fundamental and oft-cited work in this context is Émile Durkheim's *The Elementary Forms of Religious Life* ([1912] 2008), in which he explores how these forms "permeate not only traditional but modern societies as well" ([1912] 2008: vii). Durkheim argues that rites are equally crucial as beliefs in constituting religion and religious communities, and social action and social communities more generally ([1912] 2008: 36). As the religious studies scholar Catherine Bell observes in her instructive introduction to ritual theory, in the first two parts of his book Durkheim develops an analytical perspective that considers belief (thought) to precede and inform ritual (action). In the third part, however, which focuses on cult, he "reintroduces ritual as the means by which collective beliefs and ideals are simultaneously generated, experienced,

4 The description of my experience of the ceremony in the vignette is certainly no innocent description of what I observed, nor is it the only possible experience and interpretation; it has a lot to do with me being the experiencing subject. Since I was socialised in Catholic and Ecumenical Church environments, the stories, symbols, and rituals I am reminded of in such a context will most likely be those of the Christian Church. Building on the Harawayian paradigm of situated knowledges (Haraway 1988), I want to stress here that recognising this does not relativise so much as contextualise my experience and interpretation as well as the questions I deduce from it.

and affirmed as real by the community" (Bell 1992: 20).[5] The performative understanding of rituals crystallising here has become dominant in ritual studies and ritual theory since the 1970s, and has been paralleled by a broadening of the analytical focus beyond religious and spiritual rites in the narrower sense to "ritualised action" more broadly, which is also referred to as "performance" (see Bell 1992: 37–46; Belliger/Krieger 2008). These approaches aim to replace the classical dichotomic and consequentialist understanding of thought and action, theory and practice, or script and performance with a co-constitutive understanding of action and meaning (see Belliger/Krieger 2008: 9–10).[6]

An important representative of the performance approach to ritual(ised action) theory is Roy Rappaport (e.g. 1979, 1999). He proceeds from a broad definition of ritual as "the performance of more or less invariant sequences of formal acts and utterances not entirely encoded by the performers" (Rappaport 1999: 24).[7] Unlike other prominent scholars in performance theory who analyse ritual(s) by analogy with performance(s) in theatrical drama (e.g. Goffman [1967] 1989; Grimes 1982; Turner 1982, 1995), Rappaport conceives of ritual(ised action) as *performative* in the sense of performative speech, that is, as not merely

5 Bell goes on to argue that the latter perspective is closer to the understanding of ritual developed in Anthropology and at the intersection of Sociology and Anthropology, which tends to give analytical primacy to social action rather than thought (e.g. Hubert/Mauss [1898] 1981, quoted by Bell 1992: 15). More recent research on ritual in Cultural Anthropology proceeds from this perspective, which "coincides with the emergence of culture as a category of analysis" (Bell 1992: 15).

6 In a more critical reading of performance theory as basically reiterating J. L. Austin's (1975) speech act theory, which many accounts of ritual(isation) as performance proceed from, Bell (1992: 43) questions whether these approaches actually succeed in overcoming the primacy of thought over action if they ultimately understand ritual as communication. Identifying "a general tendency in the social sciences to 'textualize' the objects of its [sic] concern" (Bell 1992: 45) in this approach, she criticises the purported turn to action in ritual studies and ritual theory as potentially illusionary.

7 Rather than understanding ritual as an element of religion, Rappaport focuses on rituals' form (and what they perform). He proposes to analyse ritual/ised action on a continuum "of increasing formality and decreasing spontaneity" (1999: 34). Religious phenomena, which his study nonetheless focuses on, "lie toward the more formal, less variant end of the continuum [and involve] rituals sufficiently elaborate to include what may be called 'liturgical orders': more or less invariant sequences of formal acts and utterances of some duration repeated in specified contexts" (1999: 35). My analysis of the ritualised seed storing ceremony in Svalbard lies towards the other end of the continuum insofar as it is much less formal and more variant than rituals with a liturgical order.

saying, but doing or *enacting* something (Rappaport 1999: 107; see also Bellah 2003: 37–39).[8] His performative conception of ritualised action has a number of implications that are instructive with regard to analysing the meaning and effects of the ritualised seed storing ceremony in Svalbard described above.

First, Rappaport (1979: 177) argues, while dramas (in the classical sense) are staged for a passively consuming audience, rituals involve an actively participating congregation. Second, the acting in dramas (or plays) is not serious whereas rituals are characterised by a seriousness that is crucial for the durability and momentousness of that which is enacted beyond the confined time-space of the performance. For Rappaport, this does not mean that rituals are ultimately instrumental, though. Instead, he defines ritual as "the non-instrumental aspect or component of events that may also include an instrumental component" (Rappaport 1979: 177).[9] Third and resulting from the former, ritual performance is (potentially) transformative for those who participate in it on a deeper, more durable level than the temporary assumption of a role in a play (see ibid.).

Summarising Rappaport's argument, the sociologist of religion Robert Bellah argues that "[t]he sheer act of participating in serious rituals entails a commitment with respect to future action, at the very least with one's fellow communicants" (2003: 38). Against this background, the active involvement of the depositors participating in the event in a ritualised act of depositing their seeds in the Seed Vault can be interpreted as a means of *committing* them to the cause of agrobiodiversity conservation on a level other than that of merely rationally *convincing* someone of something. The collective ritual performatively does something to the group of depositors (and probably to other participants as well) that exceeds the instrumental or pragmatic act of backing up a seed collection in one's responsibility by depositing a duplicate collection in the Seed Vault. It performatively entangles this individual act in a collective cause or belief that is part of the fabric of the *ex situ* world of conservation.

For Bellah, what is most interesting about both Rappaport's and Durkheim's sociological analyses of ritual is not only their contribution to the sociology

8 Nevertheless, Rappaport sticks to characterising ritual as "performance".

9 Bell (1992) astutely cautions against a simplistic dualist distinction between instrumental action and a non-instrumental notion of ritual as mere symbolic action. This, she argues, "can easily collapse into a distinction between the rational and the irrational or the logical and the emotional" (1992: 71) and thus reify the power relations these dualist categories are situated in.

of religion but also the insights they offer into what he calls "the ritual roots of society" (2003). Rappaport understands ritual as "*the* basic social act" (1999: 138, emph. in orig.) insofar as it does not merely symbolically represent but enacts a social contract. Invoking this assertion, Bellah argues that "the most important implications [of Rappaport's study of ritual; FV] have to do with the creation of social conventions, a moral order, a sense of the sacred, and a relationship to the cosmos" (2003: 38).

Considered through this lens, the religiously-symbolically charged ritualisation of the act of storing seed deposits performatively enacts a morally charged convention of saving endangered agrobiodiversity through *ex situ* seed conservation as well as a sacrality pertaining to both what is conserved and the act of conservation. Through the ritualised storing ceremony, agrobiodiversity conservation thus becomes something more (note: more, not other) than a pragmatic technoscientific procedure – something transcending that which the seed summit performed and is able to perform, given its form. The ceremonial storage ritual adds a layer of meaning to agrobiodiversity conservation that has to do with enacting a relationship to the cosmos, as Bellah puts it, more than with the specific, mundane why and how of conservation.[10]

I have shown in previous chapters that the more-than-human relations the *ex situ* world of conservation is built on and enacts are shaped by a "resourcist cosmovision" (Fenzi/Bonneuil 2016: 78). The cosmorelation enacted in the seed storing ceremony, in contrast, is one in which the Human assumes the role of the preserver, caretaker, and guardian of a Nature under threat. While this may seem contradictory, what becomes discernible here are different manifestations of one and the same relation to the natural/ised world, within which the latter can take on different forms but is always defined in a subordinate relation to the human. Following the ecofeminist philosopher of science Carolyn Merchant (1980), they can be read as different manifestations of a patriarchal relation to the natural/ised world, which simultaneously imagines Nature as nourishing resource (embodied by the figure of the mother), powerful threat (embodied by the figure of the witch), and powerless protégé (embodied by the figure of the virgin or girl). In chapter 3.1, I showed how this relation has come to matter throughout history in different forms of frontierism – as

10 The notion of the cosmos is not defined any further by Bellah. I understand and use the term in the anthropological sense introduced earlier, as the world/s as understood and lived according to the plurality of cosmologies existing in the world.

territorial appropriation (nation-making frontier), technologically driven resource extraction (resource or technofrontier), and selective conservationism (salvage frontier). All of these can be discerned in the history of human activity on Svalbard. In the following, I discuss how this relation comes to matter in the salvationist imaginaries associated with the Seed Vault.

"Noah's Ark for Seeds" and the "Doomsday Vault" as Secular Salvation Stories

The idea of the Svalbard Global Seed Vault as a custodian of a Nature under threat of extinction is what makes the metaphor of a Noah's Ark for seeds seem like an apt characterisation. This role, as ascribed to the Seed Vault, corresponds to the role God assigns Noah in the biblical story of Noah and the Flood as told in chapters six to nine in the Book of Genesis, the first book of the Old Testament in the Christian Bible.[11] The story of Noah takes place in the ninth generation after Adam and Eve, according to the Bible the first humans created by God, who were banished from the Garden of Eden God had created for them because they ate from the forbidden Tree of Knowledge instead of humbly shepherding the creation as God had commanded them (Genesis 1–3).[12] Nine generations later, the story continues, God sees the depravity of the human race, regrets the creation of Man, and decides to annihilate life on earth through a great Flood. Because God considers Noah a pious man, God tells him to build an Ark, which will save him from the Flood along with his wife and sons and their wives, as well as one male and one female of every animal species. Noah and those he saves survive the Flood on the Ark and repopulate the earth afterwards.

The metaphor of the Svalbard Global Seed Vault as a Noah's Ark for seeds allegorises the promise that those who devoutly honour and save creation will survive the local and global disasters of this time and age and rebuild the world with the help of the diversity saved in the Seed Vault – which is also metaphorized in Barroso's characterisation of the Seed Vault as a frozen

11 Genesis is also the first of five books of the Torah in the Hebrew Bible. In what follows, where I recount the biblical stories discussed here, I take them from the Christian Bible in the Catholic tradition, which also comprises the New Testament and therein the story of the Apocalypse in the Book of Revelation.

12 For the original biblical stories quoted here and throughout this chapter, see https://www.bible.com/bible/ (last accessed July 18, 2025).

Garden of Eden (see above). Although these metaphors themselves were repeatedly rejected throughout the 2020 Svalbard Seed Summit, their promissory message was echoed in the story of the Syrian crop diversity rescue mission (see the book's introduction), which was not left untold as the most salient example of the Seed Vault's value. Even more significant in terms of reproducing salvationist imaginaries, however, was the seed storing ceremony, which seemed to embrace and invoke the biblical stories and metaphors – on the one hand through religious and maritime symbolism and on the other hand by performatively enacting a humility before creation, which in the biblical story of Noah and the Flood prompts God to save Noah and his kind from his wrath. Thus interpreted, the seed storing ceremony appears to be an enactment of a secular story of salvation. "Secular salvation history", according to Haraway (1997: 10), "depends on the power of images and the temporality of ultimate threats and promises to contain the heteroglossia and flux of events." Put differently, secular promises of salvation need powerful symbols and apocalyptic stories in the same way as religions do, to mobilise hope and trust in the promised means of salvation (see also Haraway 1997: 41). The most prevalent and powerful apocalyptic story told around the Seed Vault is the story of a "doomsday vault", which references the biblical story of the Apocalypse.

The Apocalypse (from Ancient Greek: *apokálupsis* for "disclosure") is the subject of the Book of Revelation, the last book of the New Testament in the Christian Bible. It takes the form of a letter written by John of Patmos to seven nascent congregations in the province of Asia (see Keller 2021: viii). Encompassing twenty-two chapters, this story is much longer than the previously mentioned ones, which are restricted to a few chapters within the Book of Genesis. Very briefly summarised: Through a series of prophetic visions, John announces the coming of God, whose wrath will descend on the earth in the shape of several dreadful creatures and cataclysms destroying "those people who destroy the earth" (Revelation 11:18). At the end of the book, John prophesies the coming of "a new heaven and a new earth" (Revelation 21:1) awaiting those who have remained humble and faithful. Alluding to this story, the notion of the doomsday vault construes the Seed Vault as a protective bunker for the world's plant genetic resources securing them from whatever apocalyptic catastrophe might befall the earth.

It is noteworthy that the first two metaphors, the Seed Vault as a frozen Garden of Eden and a Noah's Ark for seeds, reference the first book of the Christian Bible (Genesis) whereas the doomsday narrative references the last book (Revelation). In the dramaturgy of the Bible, the two books, which tell of the creation and its demise, narratively represent the Alpha and the Omega – which is also a cipher for God. In this sense, according to the feminist theologian Catherine Keller, they do not mark and tell the beginning and the end of a linear story. Rather:

> "The God who had spoken in the first chapter to say 'I am the Alpha and the Omega' now speaks again in the end: 'I am the Alpha and the Omega, the first and the last, the beginning and the end.' So in this insistently nonlinear temporality, Genesis and Apocalypse seem to name the two edges of one being, one becoming, one process of creation. But the first and the last do not become the same. They *coincide*. At its omega point the creativity of genesis (the Greek word for 'becoming') reveals itself as *re*genesis." (Keller 2021: 172, emph. in orig., quoting Revelation 22:13)

It is in this insightful remark that I recognise one of the crucial ways in which it matters that the story of the Seed Vault is told with these (and not other) biblical stories: where the seeds of the disappearing world become the seeds of a reappearing world, the biblical metaphors of Genesis and Apocalypse become the perfect frame narrative for the Seed Vault's purpose and promise. In what follows, I interrogate the doomsday narrative in the story of the Seed Vault through the biblical story of the Apocalypse, conceiving the latter as an allegory of contemporary global crises. In doing so, I follow Keller's allegorical approach to contemporary apocalypticisms, which she develops in her monograph *Facing Apocalypse: Climate, Democracy, and Other Last Chances* (2021; see also 1996).

Keller unfolds a demystifying reading of the biblical story of the Apocalypse that counters literalist interpretations by historicising the story. This approach offers valuable insights for a sociological engagement with contemporary apocalyptic storytelling, which is currently thriving vis-à-vis ever-escalating global social-ecological threats. Keller discusses the biblical Apocalypse as a historically situated prophetic parable, not in the sense that it predicts contemporary global crises but in that it has analytic value due to its potential to reveal fatal transhistorical patterns:

> "The ancient text reads the crisis of its own historical context. It does not know or predict ours. But it discerns certain patterns in its own world deep enough to persist, dangerously, and perhaps disclosively, into our own. To

> *mind* those patterns without literalizing them means to dreamread collective crisis now, by way of the metaphors – the metaforce – of the Apocalypse then." (Keller 2021: 3, emph. in orig.; see also xiii)[13]

Based on her method of "dreamreading" contemporary crises through persisting deep historical patterns, Keller proposes an analytical strategy she calls "apocalyptic mindfulness" (2021: 3). It allows for mobilising "the metaphors – the metaforce" of the Apocalypse without succumbing to the more common apocalyptic sentiments of "optimistic denialism and pessimistic nihilism" (2021: 17; see also Haraway 2016: 3–4). For Keller, being mindful of the Apocalypse means "fac[ing] the unspeakable catastrophes that may become inevitable *if* we do not speak" (Keller 2021: 17, emph. in orig.). It means taking seriously the persistence of apocalypticism (in societies with a Christian tradition or influence), recognising its various overt and covert, both spiritual and secular forms, and neither literalising nor ridiculing them (see Keller 2021: xii–xiii, 15–19).

Apocalyptic mindfulness, thus, allows for becoming aware of the cultural force that the biblical text has to this day, not only where it is explicitly referenced but also as an interpretive frame of reference in "the political unconscious" (Keller 2021: 18, quoting Jameson 1981). This is crucial not least because, as Keller argues, it is where apocalyptic sentiments are unconsciously enacted that they are most dangerous. Minding the Apocalypse and bringing to the surface its subliminal metaforce in technoscientific salvation stories and projects, then, makes it possible to disentangle apocalyptic narratives and sentiments from the forces that doom specific worlds, and thus to develop meaningful responses to the latter (see also Keller 2021: xii).[14]

---

13 Keller emphatically distinguishes between prophecy and prediction, defining the former as the analysis of "potent pattern[s] of human civilization" (2021: xiv), which may or may not repeat themselves throughout history. In this sense, "[p]rophecy then or now *dreamreads* a collective context. Within the patterns of what has already become, it attends to what might yet be." (Keller 2021: xiv–xv, emph. in orig.)

14 Keller's approach to facing the crises of the present through apocalyptic mindfulness resonates, to a certain degree, with Haraway's call to stay with the trouble(s) of our times rather than losing track of the present through abstract futurisms. Although Haraway, unlike Keller, is not interested in apocalypticism, which she conceives as the equally passivating counterpart to the contemporary "comic faith in technofixes, whether secular or religious" (Haraway 2016: 3), very like Keller, and as unfolded in chapter 4.2, she seeks and calls for responses to the devastations of the present that

An important observation Keller makes in this regard is that today, apocalypticism is no longer restricted to religious literalist traditions and the religious-political wing of reactionary conservatism, as was long the case (at least in the US context, which her research focuses on; see Keller 1996; see also 2021: 7). Instead, varieties of apocalypticism permeate contemporary crisis discourses across all kinds of societal groups and movements, whether religious or political, reactionary or revolutionary (Keller 2021: xiii). This observation is also reflected in a growing academic discourse on contemporary apocalypticisms. Scholars from a variety of disciplines and research fields[15] have been developing an ever-expanding index of the ways in which apocalyptic and post-apocalyptic narratives, sentiments, politics, and cultures shape climate, environmental, nuclear, and other existential discourses and struggles of the present. For Keller, what is particularly noteworthy in the contemporary inflation of apocalypticisms is the "mounting chorus of voices secular, sober and scientifically tuned who make use of the rhetoric of the apocalypse" (2021: 9; see also Riesch 2021). Among these are alarmist voices rather explicitly predicting the Last Days, such as "'the Insect Armageddon,' 'the Uninhabitable Earth,' 'Climate Doom' – and of course the 'Anthropocene Apocalypse'" (Keller 2021: viii). In other cases, the apocalypticism is more implicit and ambiguous. The popular story of the doomsday vault ties in with the first type of doomism. While the official story of the Seed Vault rejects this perception, it does nonetheless exemplify the second, more subtle type of apocalypticism, as I show in the following. It is here that I recognise the complicity of the partners behind the Seed Vault in the lack of success in dispelling the doomsday narrative.

The crux is that defensive statements arguing that the Seed Vault is "not about the end of the world" (Dempewolf 2020; see above) presume and reject a very particular understanding of 'doomsday' as a singular apocalyptic event in the (varyingly distant) future that will bring *the* world to *an* end. This conception

---

lie beyond the simplistic alternative of techno-optimism and "bitter cynicism" arising from "a position that the game is over" (ibid.).

15 These range far beyond theological research areas to sociology (e.g. Bröckling 2023), philosophy (e.g. Žižek 2022), anthropology (e.g. Danowski/Viveiros de Castro 2017), environmental humanities (e.g. Higgins 2021), postcolonial international relations (e.g. Mitchell/Chaudhury 2020), affect studies (e.g. Grillmayr/Hentschel 2024), and feminist theory (e.g. Zylinska 2018).

has three implications worth problematising. First, understanding the apocalypse as a singular and all-encompassing destructive event universalises and dehistoricises both *the* world and *the* assumed end awaiting it. For one thing, doing this obfuscates and depoliticises the powers that are bringing about the anticipated end. For another, it fails to see that there are worlds in this world that have already experienced or are currently experiencing the devastation or end of their known worlds. To put it in the words of philosopher Marina Garcés: "assum[ing] that the end of one world is the end of *the* world is a terrible way of forgetting that this world we now see shaking has risen on the destruction of others and on much death, as is happening again" (Garcés/Hentschel 2023: 1, emph. in orig.; see also Whyte 2017; Yusoff 2018).

Among the most cataclysmic examples of this in recorded human history was the eradication of around 95 per cent of the Indigenous population of the Americas during the two centuries following the arrival of Columbus in the Caribbean in 1492 and the ensuing colonialist destruction of the world they had previously cultivated and lived in (Danowski/Viveiros de Castro 2017: 104–105; see also Lewis/Maslin 2015: 174–175). According to philosopher Déborah Danowski and anthropologist Eduardo Viveiros de Castro, while the European colonisers conceived the American continent "as a *world without humans* – be it because they objectively depopulated it or be it because the humans they found there did not fit the category of 'humans' – the surviving Indians [sic] [...] found themselves as *humans without world*" (2017: 105, emph. in orig.). They were forced to learn how to survive the end of their world and adapt to the new world imposed on them. Insofar as this was but one of a horrific diversity of historical and contemporary examples of worlds' ends, Danowski and Viveiros de Castro describe "the end of the world [as] a truly *fractal event*" (2017: 105, emph. in orig.).

Second, understanding the A/apocalypse[16] as 'the end of the world' also dehistoricises and depoliticises the biblical story of the Apocalypse, based on (and reproducing) a simplistic literalist understanding of the Bible that political theologians such as Keller firmly oppose. According to her reading of the

16 Keller (2021: 2) distinguishes between the biblical Apocalypse and the apocalypse as a trope in current crisis metaphors by capitalising the former and lowercasing the latter. She also introduces the notation "A/apocalypse" (2021: 7) for arguments referring to both stories at once. I follow her in both the distinction and the conjunction of the Apocalypse (biblical story) and the apocalypse (trope) insofar as this notation emphatically illustrates the argument I develop in this chapter.

Apocalypse, which resonates with the first point of criticism of the 'end of the world' narrative, the story the Book of Revelation tells neither conceives the catastrophe as singular nor the world or its end as absolute:

> "John's letter literally, to the letter (Lat. *littera*), does *not* announce 'the end of the world.' It does depict a symbolically supercharged spiral of catastrophes, vividly amping up the destruction of a *particular* world: a specified global civilization and its planetary ecology." (2021: 6, emph. in orig.)

The popular idea of the Apocalypse as the story of a more or less exogenous shock brought upon the world as a whole by a chastising God discontent with humanity as a whole, then, obscures a tone in John's letter Keller reads as "profoundly critical of the empire that had colonized what for him counted as the known world" (2021: viii). John's prophecy, she argues, anticipates a backlash against the Roman Empire's violence exerted through colonialism, imperialism, and economic injustice, as well as against the more-than-human world (2021: 30). Insofar his letter is addressed to a readership living under the rule of the empire he criticises, this criticism is encoded in an ancient symbolism, which makes it less evident to modern readers (Keller 2021: xix).

These first two problematisations show that interpreting the story of the A/apocalypse as being about a singular doomsday that amounts to 'the end of the world' is based on a double depoliticisation: it depoliticises historical and contemporary forces of doom by invoking a religious narrative that is itself depoliticised insofar as it is treated as a myth rather than a historically situated story of a specific, destructive and therefore doomed world. This idea of the A/apocalypse as a doomsday myth that has nothing to do with real world problems underlies the rejection of the 'end of the world' narrative in the story of the Seed Vault. As Fowler puts it,

> "[t]he major threats and the principal causes of loss of diversity in genebanks have to do with institution-specific management, infrastructure, and funding problems. They are not catastrophic or apocalyptic; they are not the stuff of newspaper headlines. But they are deadly nonetheless." (Fowler 2008b: 12)

At the same time, stories like that of the Syrian seed collection, lost in the course of the Syrian civil war and rebuilt from the duplicates stored in Svalbard (see introduction), or that of the Potato Park, collaborating with the

Seed Vault to secure their unique biocultural heritage against the threats climate change poses to the Andean social ecology (see chapter 4.2), show that agrobiodiversity loss is connected not only to mundane problems but also to catastrophic and, for those affected, potentially apocalyptic scenarios as well. More in line with this, Fowler argues elsewhere:

> "I'd say doomsday is happening everyday for crop varieties [...]. Lots of people think that this vault is waiting for doomsday before we use it. But it's really a backup plan for seeds and crops. We are losing seed diversity every day and this is the insurance policy for that." (Fowler quoted in Handwerk 2012; see also Wallace 2016: 115–116)

Both arguments quoted here show that Fowler *rejects* the doomsday narrative based on a notion of the A/apocalypse that corresponds more to the popular imagination of a singular and somewhat exogenous apocalyptic event than to the more politicised version of the story unfolded above. At the same time, the idea he *articulates*, namely that doomsday is an everyday fate for crops, distinctly resonates with Keller's above-quoted description of the A/apocalypse as a "spiral of catastrophes" (2021: 6) emerging from the world they threaten to destroy. However, these catastrophes (in the biblical story: war, famine, pestilence, and death, symbolised by the four horsemen of the Apocalypse riding the earth one after another) do not wipe out *the* world all at once but reach apocalyptic magnitude only in their accumulation and intersection. Similarly, the everyday threats that natural and man-made disasters as well as more mundane technological, infrastructural, and financial limits pose to plant genetic diversity result in extinction-level events and the loss of an existentially critical mass of agrobiodiversity only when local losses globally accumulate and intersect in their consequences. The everyday doomsday for crops Fowler describes, then, is not at all dissimilar to the biblical Apocalypse. Both are distributed not only spatially but also temporally; or as the literary scholar Molly Wallace puts it, they are "at once future and present" (2016: 115).

This brings me to the third problematisation of the popular 'end of the world' narrative. It is an expression of a world whose imagination of a contemporary apocalypse is shaped by the twentieth century nuclear age. The nuclear threat is that of a "fast apocalypse" (Wallace 2016: 7) occurring on a specific dooms*day*. The ecological dooms*days* (in plural) that have arisen on the horizon of the twenty-first century, however, are better described as a "slow apocalypse" (Wallace 2016: 17, quoting McMurry 1996). Although Wallace barely elaborates the notion of the slow apocalypse, it can be read in line with a body of literature elaborating on the slowness of violence (Nixon 2011; Masco 2017), disaster (Knowles 2014), and emergencies (Anderson et al. 2020). All these approaches challenge an understanding of violence, disaster, and emergencies as something extraordinary, instantaneous, and immediately perceptible. Instead, they emphasise their ordinariness, temporal and spatial dispersal, and therefore imperceptible, incremental unfolding.

This perspective is crucial insofar as the slowness of apocalyptic social-ecological threats such as large-scale (agro)biodiversity loss and climate change may make them seem less dangerous than the nuclear threat because there is still time to act. However, it is precisely their slowness that makes the social-ecological apocalypses of the present all the more dangerous. As the anthropologist and STS scholar Joseph Masco puts it, they exert "a slower violence that is treacherous precisely because it is so incremental that it is difficult in any given moment to sense a change in the environment or to connect discreet issues" (2017: 70). According to Wallace, the iconic metaphor of the "doomsday clock" of the Bulletin of the Atomic Scientists (2025) ticking towards midnight as the moment of nuclear annihilation

> "might better be conceived, then, not only as a countdown to some potential future annihilation, but also as [...] a 'metronome of threat,' syncopating a present that contains multiple catastrophes, historical and to come, simultaneously. In the second nuclear age, we seem to find ourselves inhabiting rather than anticipating the end" (Wallace 2016: 17, quoting van Wyck 2005: 20).[17]

In summary, becoming aware of the slowness of the A/apocalyptic is crucial because it unfolds in multiple spatially and temporally distributed forms that can make the apocalyptic effects of their intersection indiscernible.

Following Keller's reading of the A/apocalypse, recognising the slowness and spatio-temporal dispersal of the A/apocalypse is also essential because it

17 This distinction must be read as a genealogical typification of apocalypses rather than a sequential development, as not least the current trend towards militarisation and warfare as a response to international geopolitical conflicts underlines. Even more, insofar as the competition for access to energy sources and other natural resources is one of the (surprisingly little debated) drivers of current wars and territorial conflicts (see e.g. Sánchez Cedillo 2023), the divergent existential threats of the present must be considered as intersecting and the contemporary apocalyptic threat as multiple.

opens up "a space where late chances, last chances, remain nonetheless real chances" (Keller 2021: xiii). While popular accounts of the A/apocalypse usually focus on its destructivity, Keller highlights the "*constructive* potential" (2021: 9, emph. in orig.) inherent in the alarm of the A/apocalypse. She argues that this is a core message of the Book of Revelation: "*Apokalypsis*, the Greek word for 'revelation,' means literally the 'removal of the veil.' It means not closure but dis-closure – that is, opening." (2021: 5) Without questioning the fact that the story of the Apocalypse incontrovertibly tells of the destruction of a world, Keller points out that "contrary to general presumption and Christian literalism [...] neither the book nor its world ends there" (2021: 6). As mentioned above, John's prophecy, while by and large indulging in visions of devastation, is not restricted to these. It ends with a promise of salvation and a new world awaiting the faithful, which refers to those "who [do not] destroy the earth" (Revelation 11:13; see above). This revelation of a new world to come, Keller argues, comes with a hope that "eschatologically tuned" (2021: 8) progressive movements – historically Christian ones but today including secular movements – have translated into visions and demands of "social transformation" (ibid.). Crucially though, she points out, "for any biblically sound eschatology, it is not that something new will 'take the place of the old'; 'it is the same "old" itself which is going to be created anew'" (Keller 2021: 8, quoting Moltmann 1996: 88).[18]

Against this backdrop, the Seed Vault's promise to safeguard the largest possible diversity of plant genetic resources in order to open up options for adapting agriculture to unknown future ecologies and continuously (re)creating an abundance of thriving food crops reads like a hopeful promise of salvation in the face of a slowly unfolding social-ecological apocalypse. Although the apocalyptic tone I recognise in the rationality of *ex situ* conservation encapsulated in this promise certainly has little to do with biblical literalism, it does conform to a type of apocalypticism Keller distinguishes from the former, which she calls "modern *'can do' optimism*" (2021: 16, emph. in orig.).[19] The understanding of the A/apocalypse that modern 'can do' optimism enacts embraces

18 Following the theologian Jürgen Moltmann, Keller makes a similar argument with regard to the tradition of eschatology as she does about the Apocalypse. She argues that it has "misleadingly [been] read as 'doctrine of end things,' [whereas it] is about a hope that is 'revolutionizing and transforming the present'" (Keller 2021: 8, quoting Moltmann 1967: 16).

19 Keller's (2021: 16–17) typification of contemporary apocalypticisms encompasses four variants: biblical literalism, catastrophist nihilism, its flipside: modern 'can do' opti-

the constructive potential of John's revelation through the belief in human exceptionalism, technological progress, and economic growth as the means to save the modern world from doom (note: the modern world, not *the* world).

Through the promise that the Seed Vault will make it possible for creation to survive the coming catastrophes, this mode of resisting the destructivity of the A/apocalypse also promises to avert the need for any radical transformation of the social (human and more-than-human) relations from which these catastrophes emerge. The 'can do' optimism that informs this promise is inscribed into the conviction that conserving plant genetic resources *ex situ* can do what cultivating agrobiodiversity *in situ* cannot do, namely save as much plant diversity as possible from going extinct while also organising global agriculture in a way that is able to feed the entire world population (irrespective of whether it does or not). It is further inscribed into the conviction that the Seed Vault can do what local genebanks cannot do on their own, namely guarantee the survival of plant genetic diversity and thus secure nothing less than "the future of agriculture" (Fowler 2008b) amidst the many catastrophes and local losses of diversity happening every day. Finally, it is also inscribed into the religious aesthetic and ritualised character of the seed storing ceremony conducted in Svalbard in February 2020, which, against this background, appears as a celebratory enactment of the salvationist promises of and belief in putting seeds on ice to survive the looming agrobiodiversity doomsdays.

What I have attempted to show throughout this discussion is that although the Svalbard Global Seed Vault is not "the doomsday vault [preserving] the seeds that could save a post-apocalyptic world" (The Guardian 2015) that popular media accounts persistently make it out to be, it does embody a promise to its depositors that it will help them survive their own respective doomsdays. Ritualised events such as the seed storing ceremony analysed above enact and, by doing so, commit its participants to the secular salvation story of the Seed Vault. Despite the persistent rejection of the doomsday narrative, the Seed Vault's secular promise of salvation is infused with a quite biblical apocalypticism insofar as it is the conservation of creation and the belief that lost worlds can be recreated that underlies the promise of salvation from doom. This type of apocalypticism does not address the forces of doom, assuming that they lie outside of anyone's sphere of responsibility or influence. Instead, hope emanates from the belief that the regeneration of creation as securely

---

mism, and the apocalyptic mindfulness she proposes as a strategy that resists the fallacies of the other three.

conserved in the Ark (be it the biblical one or the Seed Vault) will enable it to survive the coming catastrophes. In the secular technoscientific context of the *ex situ* world of conservation, this apocalypticism takes the form of a technoscientific 'can do' optimism that comes with a depoliticisation of contemporary agrobiodiversity doomsdays.

Considering *ex situ* conservation practices and the Seed Vault not in themselves, but as performative parts of the *ex situ* world and its mode of world-making, makes this depoliticisation all the more notable. As I have argued in previous chapters, the *ex situ*ating mode of world-making that the *ex situ* world conserves and that the Svalbard Global Seed Vault reinsures seeks to extend the present of a world valuing and striving for a form of technoscientific and economic progress that is built on exploiting nonhuman/ised life as a resource available and extractible for the flourishing of that world. Insofar as this world and its mode of world-making are complicit in the slow ecological apocalypse that has appeared on the horizon of the twenty-first century, saving it means saving its apocalypse, too. The way in which the Seed Vault and the *ex situ* world of conservation currently act on the space of late and last chances unveiled by the apocalyptic threats of the present, then, is a mode of worlding trapped in a vicious cycle of loss and salvation.

Nevertheless, some of the other worlds that the Seed Vault assembles show that other modes of knowing, making, and being in the world exist and persist in spite of loss and destruction. Based on my analytical approach to the Seed Vault as an arena of world-making that is not determined by the world it has arisen from and following Danowski and Viveiros de Castro's appeal that other worlds "have something to teach us when it comes to apocalypses, losses of world, [...] and ends of History" (2017: 104), the section that follows shifts the focus to other responses to the contemporary social-ecological catastrophes emerging in the space of the late and last chances they unveil.

5.2 Conservation as Commemoration: The Art of Mourning Agrobiodiversity Loss

> "Longyearbyen is home to the most biodiverse room in the world. Housing over 1 million seed samples from all around the world, the Svalbard Global Seed Vault is an incredible achievement of science and diplomacy. However, the Vault contains no information about the cultural significance of seeds; no stories about how they are cultivated, by whom, for what purposes, using

> what rituals, etc. The seeds in the Vault are frozen in isolation from all the social practices, ecological relations and cultural histories that give them life. The Agri/Cultures.Seed-Links exhibition is a reminder that seeds live within rich webs of human/nature relations and celebrates these connections as also worthy of preservation." (GenØk – Centre for Biosafety 2019)

The passage quoted here comes from a poster announcing the *Agri/Cultures.Seed-Links Exhibition: Conserving Cultural Connections with Seeds*, staged in Svalbard on June 8, 2019, at Huset, a conference venue just outside Longyearbyen. I happened to come across this art exhibition, which lasted for this one day only, during my first field research trip to Svalbard. It was part of the public outreach of the Agri/Cultures Project, a research project funded by the Norwegian Research Council aiming to create publicly accessible knowledge about the diversity of agri-food systems and cultures of agriculture.[20] In line with this goal, the exhibition in Svalbard was curated around the cultural contexts of seed life that *ex situ* agrobiodiversity conservation efforts deprioritise. As the text on the poster announcing the exhibition illustrates, it is a direct response to the Svalbard Global Seed Vault and the *ex situ* approach to agrobiodiversity conservation, as well as to the shortcomings of this kind of conservation as identified by the Agri/Cultures Project. This art project represents one of the actors and activities the Seed Vault less evidently assembles in the political world-making arena it has created. This section shows that extending the focus of analysis beyond the more obvious practices and environments of agrobiodiversity conservation to cultural and artistic interventions into the world of conservation makes it possible to articulate a critique of the *ex situ* world of agrobiodiversity conservation that is productive insofar as it draws attention to other possible worlds and modes of worlding.

To do so, I develop a twofold argument that emerges from two distinct levels of intervention I identify in the art project. Insofar as it consists of an exhibition of artworks illustrating what I call agrobiocultural diversity, on the one hand, and a performance in which these artworks are deposited inside the permafrost for long-term conservation, on the other hand, I read it as a challenge to both the *what* and the *how* of the kind of agrobiodiversity conservation

---

20 The findings of the project are published on an interactive website visualising the relations between agri-food systems and everyday food choices, which is meant help consumers make informed decisions: https://seed-links.com/ (last accessed July 18, 2025).

practised by the Seed Vault and the *ex situ* world of conservation. First, by assembling, exhibiting, and conserving a collection of agrobiocultural diversity alongside the Seed Vault, the art project draws attention to the endangered "social practices, ecological relations and cultural histories" (see above) that come with seed diversity and from which it comes. Second, by depositing the collection through the performance of what appears to be a burial rather than a conservation procedure, which comes with a different set of affects and imaginaries, the art project allows for rethinking how to deal and learn to live with past, present, and future losses of agrobiodiversity.

Insofar as the first argument deeply resonates with the discussion of the Potato Park in chapter 4.2, I develop it rather briefly here through some of the artworks assembled in the exhibition. I then go on to unfold the second argument through a film shown during the exhibition, which documents a first iteration of the art project carried out the year before. The documentary provides deeper insights into the act and affective context of burying the artworks in the mountain than I was able to get during my visit of the *Agri/Cultures.Seed-Links Exhibition* in 2019. I interpret the performative act of burying the collection of artworks in the permafrost mountain, through the affective atmosphere it creates, as a practice of mourning the loss of agrobiocultural diversity. Based on a discussion of a body of scholarship on mourning beyond the human that has emerged in the wake of the vast ecological devastations and losses of the present, I develop a reading of the art project as a productive critique of *ex situ* conservation; one that performatively opens up new perspectives for rethinking conservation as a practice of becoming with loss rather than conserving the seeds of a world that perpetuates ecological destruction and agrobiodiversity loss.

The Agri/Cultures.Seed-Links Exhibition: Assembling Agrobiocultural Diversity

The Agri/Cultures Project and the *Agri/Cultures.Seed-Links Exhibition* both spring from a central line of criticism voiced against the *ex situ* approach to agrobiodiversity conservation and the Svalbard Global Seed Vault as its international flagship. While decades of incessant efforts to collect and conserve the largest possible amount of plant genetic diversity in genebanks all over the world and 16 years of extra security provided by the Seed Vault have successfully secured a diversity of PGRFA from loss and destruction, agrobiodiversity loss and destruction remain global social-ecological problems that have not become less existential in any way. Critics of the *ex situ* approach to conservation and of its hegemonic status within the larger world of agrobiodiversity conservation attribute the inability of *ex situ* conservation efforts to alleviate ongoing agrobiodiversity loss and destruction to, among other things, the bio-centred understanding of agricultural biodiversity underlying *ex situ* conservation efforts.

This conception of biodiversity is narrow insofar as it reduces plants and seeds to biological lifeforms and ignores the diverse biosocial and biocultural relations within which plants and seeds come to matter (see e.g. van Dooren 2009a, 2009b; Harrison 2017; Boschen 2022; and critically Laboissière 2019).[21] To borrow a terminological differentiation introduced by the anthropologist Stefan Helmreich, the notion of biodiversity and the *ex situ* approach to biodiversity conservation primarily focus on a diversity of "*life forms* – embodied bits of vitality like organisms and species" (Helmreich 2011: 673, emph. added), while disregarding the diversity of "*forms of life* – social, symbolic, and pragmatic ways of thinking and acting that organize human communities" (ibid., emph. added). In addition, as I have shown in previous chapters, by preserving a diversity of life forms in order to retain the biological and associated economic potential recognised in them, the *ex situ* world of conservation not only reduces agrobiocultural diversity to biological diversity, but in fact preserves a particular, *ex situ*ating, resourcist, capitalist form of (more-than-human) life.

The *Agri/Cultures.Seed-Links Exhibition* responded to this narrow conception of agrobiodiversity by assembling a range of artworks displaying an intriguing diversity of biocultural practices, knowledges, and contexts, in which seeds and plants exist and persist. Examples that particularly caught my attention were a photo gallery illustrating the current reality of biodiversity loss and other post-war challenges farmers are facing in northern Iraq, part of the region called The Fertile Crescent where agriculture first developed (Laurent 2019); a foldout life-sized illustration of a crocodile killed in the nineteenth century and exhibited in a museum in Singapore, in whose carcass taxidermists found a preserved wheat seed, which the artists interpreted

21 Social scientists and cultural studies scholars use the notions of the "biosocial" (Rabinow 1996; Gibbon/Novas 2008) and "biocultural" (Frost 2016) to describe the interrelatedness or hybridity of biological and social or cultural life and to dissolve reifying dualisms like nature and culture. In this chapter, I take up the notion of the biocultural rather than the biosocial following the focus of the *Agri/Cultures.Seed-Links Exhibition* on the diversity of cultural relations with seeds.

as a metaphor for entangled colonial histories of violent more-than-human relations (The Migrant Ecologies Project 2019); and a leaflet in the shape of the Seed Vault's exterior entrance structure made of hand-crafted paper bags each filled with the seeds of the plant it was made of, which the artist duo saved in their own urban gardens (Seeds InService 2019), conserved alongside a copy of the self-created book *An Illuminated Feminist Seed Bank* illustrating the traditional gardening practices and knowledge the artists engage (Potter/Puckett 2019).[22] All of these artworks illustrate an agrobiocultural diversity that is not reducible to the seeds extracted from these diverse entanglements and that is not represented, let alone conserved, in *ex situ* conservation environments, least of all in the Svalbard Global Seed Vault.

In contrast to scholarly criticism of *ex situ* conservation emphasising the limited scope of a genebanking approach to the conservation of agricultural diversity "in any full sense of the term" (van Dooren 2009a: 373),[23] as an art project with more creative options than are generally open to technoscientific projects the *Agri/Cultures. Seed-Links Exhibition* ventured to embrace the *ex situ* approach to conservation and extend it to agrobiocultural diversity. Rather than merely challenging the Seed Vault's bio-reductionist praxis of collecting the seeds of agriculturally relevant plants by assembling a counter-collection of a diversity of agricultural practices, knowledges, and traditions, the exhibition pursued a dialogical approach and offered the Seed Vault a biocultural addendum to its seed collection.[24]

At the end of the single day that the artworks were on display, as a last item on the exhibition's agenda, the artists were asked to pack their artworks into storage boxes such as the ones used in the Seed Vault. After a silent, solemn, and somewhat mournful packing ritual, the group went the next day to deposit the collection of agrobiocultural diversity alongside the Seed Vault's seed

---

22 Pictures of all the artworks assembled in the exhibition as well as video interviews with some of the artists, in which they elaborate on their exhibits and their work more generally, are accessible through an online "Seed Cultures Archive": https://www.seedcultures.com/svalbard-2019#/2019/ (last accessed July 18, 2025).

23 Van Dooren (2009a) argues that genebanks should be reimagined as crucial components of larger networks of growing, circulating, and sharing rather than conserving plant diversity in the current geneticised sense of the endeavour. Sheryl Breen (2015) unfolds a similar argument in her contrasting study of Native American seed saving projects and the Svalbard Global Seed Vault (see also chapter 2.1).

24 As part of a literal dialogue, the Seed Vault coordinator contributed a presentation about the Seed Vault to the exhibition's accompanying programme.

collection. The boxed artworks were deposited inside one of Longyearbyen's disused coalmines near the Seed Vault – whose management did not agree to storing them inside the vault. Inside the mine, under 150 metres of rock and permafrost, they now sit in a room referred to as Fröyhall (old Norwegian for "room for seeds"), which also houses the container that served as the Nordic countries' backup seed repository before the Global Seed Vault was built. "There they will rest forever, archived as cultural memories of the forgotten stories linked to these frozen seeds" (GenØk – Centre for Biosafety 2019), as the exhibition poster states.

This last passage from the poster is indicative of a crucial difference between the art project's co-collection of agrobiocultural diversity and the Seed Vault's seed collection regarding not only what is conserved, but also when and how the conserved collections come into effect. The Seed Vault promises to securely store backups of working collections of seed diversity that can be retrieved from Svalbard at any time if they are lost in the original genebank. Against this background, it works like a bank or an insurance policy rather than constituting an eternal archive in the sense of a library of seeds.[25] As I have discussed in previous chapters, while the Seed Vault is built to endure as far into the future as possible, its goal is not to sustain biological memories for eternity (or for after 'the end of the world'; see section 5.1), but to conserve plant genetic diversity for use (see chapter 4.1). The diversity conserved in the Seed Vault is not there to be remembered, but to be retrieved and regenerated if its original source is lost. Hence, the vault's temporal orientation may be long-term, depending on when genebanks ask for their deposits back, but it is emphatically this-worldly.

The frozen art collection, in contrast, expressly establishes an archive of agrobiocultural diversity that is not meant to be retrieved in the here and now. While there may be a future in which the "cultural memories" archived in this collection will be recollected, they are not deposited for any kind of anticipated

25 In contradistinction to this claim, Peres (2016), Harrison (2017), Harrison et al. (2020), and Grigoli (2024) analyse the Seed Vault as an archive of seed diversity. Insofar as they conceive archives as actively producing rather than merely representing the diversity they comprise, the approach that these authors pursue resonates with the claim that the Seed Vault is a bank more than a library. According to Harrison, practices of biobanking collect and archive diversity to "generate and accumulate latent forms of biocapital" (2017: 80).

future use. Instead, the artworks are explicitly put on ice "forever" as an eternal archive of the social, ecological, and cultural relations, practices, and histories they embody.[26] However, there is something to the *act* of conservation performed by the art project that exceeds that which it conserves (in a similar way as in the seed storing ceremony described in the previous section). In what follows, I extend the focus from the artworks – the *what* of conservation – to the performativity of the art project as a whole, meaning the process of assembling, exhibiting, packing, and depositing artworks – the *how* of conservation. I interpret the latter as an act of remembering what the *ex situ* world and mode of worlding dismembers and thus as an act not primarily of conservation so much as of commemoration.

Offerings: Conserving and/as Commemorating Agrobiocultural Diversity

Part of the accompanying programme of the *Agri/Cultures. Seed-Links Exhibition* was a screening of the short film *Offerings* (Prescott 2019) documenting an art project of the same kind called the *Svalbard Ark* that took place in 2018. Like the 2019 art project, this predecessor exhibition and storage event was organised by Fern Wickson, one of the researchers in the Agri/Cultures Project, as part of her previous research project biodiverSEEDy, also funded by the Norwegian Research Council. In the first iteration in 2018, the project began with an exhibition in Tromsø in Northern Norway (and not in Longyearbyen as in 2019), followed by a journey to Svalbard where the artworks were deposited in the same place as in 2019. The documentary follows five international artists participating in the *Svalbard Ark* from the exhibition to the storing of their artworks in Longyearbyen. In addition to documenting sequences of the events and the journey, the film is a collage of interviews with the curator and the artists accompanied by images of the Arctic landscape set to tranquil music. It provides insights into the processes mentioned above preceding and following the exhibition which I was not able to gain in person during my visit to the second exhibition in Svalbard in 2019, since my schedule did not allow me to attend the storing of the artworks in the coalmine the day after the exhibition.

---

26 Of course, the Seed Cultures Archive makes the collection accessible to a certain degree beyond its storage in the Arctic permafrost and the one-day exhibition that preceded this. However, these are digital copies whereas the originals remain on ice. While this raises an interesting question about the respective relation between original and duplicate in the art and the seed collection, these are a subject for a different paper.

To discuss the ritualised and remarkably affectively charged process of exhibiting, packing, transporting, and depositing the artworks, I now shift the focus to the documentary film of the 2018 *Svalbard Ark* project. While the film is certainly not a neutral document of the art project insofar as the way it is structured, narrated, and iconically set is already an interpretation of what it depicts, the tone and atmosphere it portrays, where not explicitly stated otherwise, resonate with my experience of the 2019 exhibition. In what follows, I unfold a reading of the art project's performance as a process of sharing, celebrating, entrusting to a beyond (beyond one's own reach as well as one's own world and time), and mourning the carefully crafted works of art embodying social and cultural relations with seeds.

The film begins by elaborating the idea of the project it documents: to complement the Svalbard Global Seed Vault's collection of decontextualised frozen seeds with a collection of artworks illustrating "the beautiful life of seeds" (Wickson in Prescott 2019: 1:08), meaning the life seeds lead in their biocultural entanglements outside genebanks. As in the 2019 exhibition described above, the artworks in the first exhibition in 2018 too embody a variety of practices, knowledges, and cultural traditions as well as personal experiences associated with seeds. They range from drawings of seeds to embroideries made by Indigenous peoples depicting their history and culture and to a shrine that used to protect seeds an artist's Hungarian grandmother took with her when migrating to the USA to bring along and continue to cultivate a piece of her home (see also Seed Cultures Archive 2018).

The film shows the exhibition itself in a lively and conversational ambience. The atmosphere changes when it comes to the packing ritual at the end of the exhibition. At the moment of packing their artworks into storage boxes, viewers see the audience silently standing by while they hear one of the artists describing the extraordinary affects that come with this peculiar experience: "That this is the first and last time this work is going to be shown was humbling and had this solemnity to it or had this seriousness to it that I hadn't felt in an exhibition before." (Sara Schneckloth in Prescott 2019: 4:18) The anticipation of letting go a carefully crafted work of art exhibited for no longer than a single day and "burying" (Schneckloth in Prescott 2019: 4:33) it in a mountain, "presumably for eternity" (Schneckloth in Prescott 2019: 2:11), as the film depicts it, comes with a suspense that is contemplative and mournful rather than resembling the optimistic and hopeful suspense of the seed storing ceremony described before. The impending storage procedure appears as a burial more than a conservation procedure, as is reflected in the way the artists talk about the project as well as in their rituals shown as the film progresses.

The affective atmosphere characterising the end of the exhibition resonates intriguingly with the way some of the artists describe their experience of the Arctic environment upon arrival in Svalbard – where, in the case of the first iteration of the art project, they did not arrive until after the exhibition, then held in Tromsø. In a voice-over to a series of images of the environment around Longyearbyen, one of the artists states: "Just getting off the plane, I was practically moved to tears by the landscape; it's really stunning and so strange, it's just really strange" (Mollie Goldstrum in Prescott 2019: 5:52). Another participant comments on the scenery as follows: "There is something about this landscape also that is kind of perfect for contemplation because it's kind of a world in suspension. I mean things are alive and the earth moves and the glaciers move and the ice moves, but there's also an awareness of stillness, a kind of silence." (David Voros in Prescott 2019: 5:24)

These passages illustrate the intensely affective, contemplative, and nearly spiritual atmosphere that *Offerings* portrays. Moreover, they introduce Svalbard as the perfect environment for putting in suspension a decontextualised collection of diverse forms of more-than-biological seed life, endangered and at the same time insufficiently conserved all around the world, because Svalbard itself appears to be "a world in suspension". As a landscape that is at once "still" and "moving", at once far removed from all that is associated with culture and civilisation and marked by the remnants of past and present extractive, frontierist cultures (see chapter 3.1), Svalbard evokes an ambiguous "affective atmosphere" (Anderson 2009) characterised by both "presence and absence, [...] the definite and indefinite" (Anderson 2009: 77). Thus experienced, it invites ambiguous kinds of contemplation or suspense.

An intriguing description of the ambiguity of "[t]he suspense generated through suspension" (Hoeyer 2017: 207) can be found in an essay on the use of cryopreservation technologies in medical contexts by the STS scholar Klaus Hoeyer. The ambiguous potentiality of life suspended in cold storage, he argues, comes with a suspense that is both promising and threatening insofar as the question of whether or not the preserved biomaterial will survive and fulfil its envisioned potential remains open as long as the material rests in suspension. Through this ambiguity, Hoeyer argues, practices of suspension and the suspense they produce open up spaces – or rather spatio-temporal situations – for action, in which meanings and social relations become matters of dispute.

In his own words: "Suspense is what drives a drama, and dramas work on the social settings in which they unfold." (2017: 211)

In medical care as well as in many other contexts employing technologies of cryopreservation (see e.g. Kowal/Radin 2017; Braun et al. 2023), including the conservation of plant genetic resources, the suspense of suspension is lifted as soon as the frozen biomaterial is reanimated, whatever the outcome. In the case of the artworks embodying a collection of agrobiocultural diversity put in suspension by way of being deposited in the Arctic permafrost for ultra-long-term conservation, the suspense that comes with the act of suspension is a different one. Here, the act of suspension is not anticipatory in the sense of being oriented towards a this-worldly future in which the potentiality of the suspended is expected to be reactivated.[27] The particular kind of suspense connected to the "burying" of artworks inside the permafrost to rest there "forever" arises from the suspension of the future itself, from an anticipation of eternal suspense. This creates a different kind of drama than the hopeful and optimistic suspense characterising seed conservation and thus yields a different kind of contemplation of the social setting, or of the form of life or world within which the drama unfolds. Here, the suspense is both hopeful and mournful. It *moves* differently (affectively speaking) and therefore comes with other kinds of imaginaries and practices.

This becomes discernible, first, in the highly affective and spiritual atmosphere that the artists and the curator evoke through the way they speak about the project – which is especially noteworthy considering that the art project is part of the public outreach strategy of a scientific research project. On the bus ride to the coalmine where the artworks will be buried, the film quotes Wickson reflecting on the project as follows: "What impact does it have to bury boxes of artwork high in the Arctic and inside a mountain? It was the gesture that felt important; the reaching out to a far distant future unknown; this offering, that nature and culture are bound, that felt like enough." (Wickson in Prescott 2019: 10:26) This passage illustrates that the act of depositing artworks embodying a diversity of social and cultural relations, giving life to the biological life of seeds inside the Arctic permafrost with its capacity for long-term conservation, is not a procedure of putting something on ice for long-term conservation in the

27 This has to do, not least, with the fact that much of the value assigned to frozen seeds is economic whereas much of the value assigned to art is aesthetic. It is not that which is suspended, however, that I am interested in here so much as the act of suspension and the suspense that comes with it.

sense of conservation as practiced in the *ex situ* world. The act of "reaching out to a far distant future unknown" that Wickson describes is not an act of *conserving* a biocultural diversity considered endangered in the present so that it will not be lost and can be regenerated. It appears more like an attempt to "offer" the conviction that "nature and culture are bound" to a future world in the hope that this ontological reality, which the present world does not sufficiently recognise or conserve, might be regenerated. Insofar as it is "the gesture" that counts, for her, and the future that will decide the artworks' fate, as the artists quoted above state, the group experiences the act of burying as "a hopeful act" that needs no resolution. In other words, it is an act of collective *commemorating*, remembering relations that have been severed by the modern form of life: biological and cultural life, but also rationality and affectivity, science and art.

A second way in which the art project *moves* differently than an *ex situ* conservation procedure becomes tangible in the ritual the artists conduct before entering the mine to deposit their artworks in the permafrost mountain. The film shows the five artists standing in a row with the boxes containing their artworks lined up beside their feet in front of the coalmine they are about to enter, with Svalbard's fjord- and mountainscape behind them. One after the other, each artist reads out a personal statement they will add to the artwork inside the box, as collectively decided the night before. While three of the five statements focus on the deposited artwork and describe its meaning as intended by the respective artist, two focus more on the performative aspect and felt meaning of the art project as a whole. Two of these statements are particularly noteworthy, both buried alongside a collection of drawings. The first shows the plants and seeds in the Syrian seed collection lost in the civil war and recreated from the duplicates stored in Svalbard; the second shows early-stage germinating seeds. The two statements read as follows:

> "By burying the originals, the drawings themselves, and not their echoes, I want it to be a hopeful act, that when, if there is future contact with the images, it is one of flesh to flesh, making to seeing, material from my hand to yours. And if there is no contact and if they meet no fate other than the darkness in the mountain, the act of making remains in memory, for a while, a ripple in the experience of a life, a moment marked by wonder, gratitude, and in feeling both humbled and hopeful." (Schneckloth in Prescott 2019: 12:13)

> "I offer these drawings, an insufficient expression of all of this, to the permafrost, and to whatever befalls them. In a climate of uncertainty and unrest, may they be points of illumination to anyone to view them." (Goldstrum in Prescott 2019: 13:18)

As opposed to the others, both of these statements are more than mere additional information on the artworks; they are part of the performative enactment of depositing, burying, offering the artworks to a beyond. They illustrate that in the moment the artists are about to let go of their crafts, it is the act that matters more than the artworks themselves and their fate. The experienced meaning of the art project, previously predominantly ascribed to the artworks, now primarily lies in the act and process of making an artwork, exhibiting it for one day only, and then burying it forever in a mountain. Here, the project becomes about the performance more than about the exhibition.

The performativity of this performance is somewhat paradoxical insofar as it enacts two antithetical things at the same time. On the one hand, by expressly *burying* and not *conserving* a collection of marginalised forms of life, it performatively reiterates the reality of global agricultural cultivation and conservation, which prioritises the form and diversity of 'life itself' (see chapter 4.1) at the expense of the diversity of other (agri/cultural) forms of life. On the other hand, by performing a burial that comes with rituals of commemoration and mourning, rather than a conservation procedure that comes with the *ex situ* world's technoscientific 'can do' optimism (see section 5.1), the art project performs a different kind of drama in the ambiguous spatio-temporal situation that the practice of suspension opens up. What it performatively offers is a multi-layered "otherwise" (Meek/Morales Fontanilla 2022) of responding to agrobiodiversity loss in the broad, biocultural sense of the term. Imitating, yet subverting the practice of conservation, it offers a different mode of dealing with the loss of the diversity of life – lifeforms as well as forms of life – that rearticulates conservation as the celebration and commemoration of agrobiocultural diversity.

In what follows, I read this performative art of conservation-as-commemoration, enacted through the practice of burying a carefully crafted work of art and entrusting it not to an anticipated this-worldly future but to an afterlife that is out of one's hands and reach, as an articulation of an emergent praxis and discourse increasingly arousing scholarly attention in the context of the contemporary social-ecological transformations: the mourning of ecological losses.

Mourning More-Than-Human Loss

In Western culture, mourning the loss of a life is traditionally an individual or family affair reserved for human lives, especially the lives of relatives and close friends. The more closely related the deceased are to the mourner(s) and the more tragic the death (e.g. early, unexpected, or violent), the more accepted are the grief and an extended period of mourning. Mourning nonhuman lives is common and accepted, if at all, primarily for household pets that have lived as quasi-related members of a human family. The dying of non-pet animals, plants (inside and outside the house), and other nonhuman lives, while it can elicit an affective response, is not usually mourned (see e.g. Braun 2017). The anthropocentrism of this understanding of loss and mourning is rooted in the dominant Western idea of the human as ontologically distinct from the rest of nature, which is seen as resource rather than kin. From the assumption that the human is distinguished by (and distinct from other animals through) his conscience rather than his natural instincts arises a predominant conception of grief as a state of mind and therefore a matter of psychology.[28]

The psychological understanding of grief and mourning dominant in much of Western thought to this day goes back to Sigmund Freud. In his seminal essay "Mourning and Melancholia" ([1917] 1957), Freud first described mourning as the "normal" (Freud [1917] 1957: 243) process of grieving the loss of a loved person or abstraction (such as one's country or ideals). According to this conception, a loss disrupts a libidinal attachment which the Ego seeks to preserve or reestablish by internalising the lost other, by making the attachment a part of the self. The process of mourning allows the Ego to work through the loss, eventually let go of the attachment, and move on as a re-integrated Ego. Freud distinguished the 'successful' work of mourning from melancholia, which he conceived as the pathological state of being entirely consumed and inhibited by the loss, unable to let go and resolve the loss and the attachment. In melancholia, he argued, the narcissistic Ego turns inward and against itself, holding on to the attachment but also to the pain caused by its loss, thus remaining trapped in depressive mourning. This early understanding of mourning and

28 The gendered language is intentional insofar as this notion of the human is modelled on a male (as well as white, European, bourgeois, able-bodied, heterosexual, etcetera) prototype – an "overrepresentation" of "Man [...] as if it were the human itself" (Wynter 2003: 260).

melancholia remains influential to this day, although Freud revised the original differentiation in *The Ego and the Id* (1923). On the basis of a reconceptualisation of the self as co-constituted by the losses and mourning processes experienced in the past, he rearticulated his understanding of melancholia as a particular narcissistic form of mourning rather than a divergent and detrimental response to grief (see also Clewell 2004; Cunsolo/Landman 2017a: 8–9; Barnett 2022: 16–18).

Freud's conceptions of grief, mourning, and melancholia have shaped much contemporary psychological and popular knowledge. However, they have also been widely subjected to critical discussion throughout the past century. A psychological strand of argument highlights that the Freudian conception of mourning hinges on an Enlightenment understanding of bounded individualism (Clewell 2004) and a separation of the emotional (grieving self) from the rational (reintegrated self) (Dodds 2011). A strand of critique rooted in Gender and Queer Studies emerges from the work of Judith Butler (2004, 2020), who critically discusses "the differential allocation of grievability that decides what kind of subject is and must be grieved, and which kind of subject must not" (2004: xiv). In contradistinction to Freud, Butler understands melancholia – which they discuss, first and foremost, as a collective sentiment – as a form of "disavowed mourning" (ibid.). Disavowed mourning refers to the mourning of lives whose loss is rendered "ungrievable" (2004: 35), for example through discursive dehumanisation, which often begins well before the loss. This most notably applies to racialised groups of people whose deaths in the context of colonialism, war, and unwanted migration tend to remain publicly unacknowledged as grievable losses.[29] Reflecting on queer experiences of differentially allocated grievability, Sara Ahmed (2004: 156) further draws attention to what psychologists refer to as "disenfranchised grief" (Doka 1989; Thompson/Doka 2017), meaning grief over losses not recognised as significant losses. Queers often experience this form of grief, according to Ahmed, not because queer losses are rendered ungrievable but rather because they tend

29 Butler developed the notion of un/grievability in reflections on the USA's "war on terror" unleashed in the aftermath of the attack on the World Trade Center in New York on September 11, 2001 (Butler 2004). Sara Ahmed pointedly observes that "the distinction between lives that are grievable and ungrievable is necessary if the 'war on terrorism' is to be justified as a recovery from terror, rather than a repetition of terror. The legitimacy of the war is inferred from the 'legitimacy' of some losses over others." (Ahmed 2004: 191–192)

to be publicly unacknowledged as losses due to a lack of acknowledgement of the relationships and forms of life within which these losses occur. What both Butler and Ahmed demonstrate is that grief and mourning are not merely personal and psychological issues but embedded in socially and politically contested forms of life.

The ungrievability or lack of public acknowledgment for losses of racialised, queer, and other marginalised lives illustrates the differential valuation of lives within the dominant form of life in Western modernity. This is also reflected in a related body of post-, decolonial, and Indigenous scholarship highlighting the ways in which egocentric and anthropocentric notions of loss, grief, and mourning are situated within "Western modernity and its colonial matrix of necropower" (Molina Vargas/Marambio/Lykke 2020: 186; see also Mbembe 2019). These authors criticise "the coloniality of mourning" (Molina Vargas/Marambio/Lykke 2020: 188) that confines grief over the loss of other forms of life to a "white humanist melancholia" (Molina Vargas/Marambio/Lykke 2020: 197), which nostalgically grieves for the disappearance of cultural diversity. This form of melancholia, they argue, ignores the necropolitics this de-diversification is entangled in and, by doing so, continues to deny non-hegemonic worlds (e.g. Indigenous and postcolonial worlds) "access to futurity" (2020: 188). This critique comes with a call for decolonising mourning by "reclaiming [...] indigenous rights and philosophies of death and mourning" (Molina Vargas/Marambio/Lykke 2020: 186; see also Braun 2017; Cunsolo 2017) and by practicing other, marginalised, resistant forms of "communal" (Muñoz 1999) or "decolonial mourning" (Gutiérrez Rodriguez 2023).

In the past two decades, a multi-disciplinary debate has arisen in conversation with these critical lines of engagement with loss, grief, mourning, and melancholia beyond the disciplinary confines of psychology that discusses questions of grief and mourning with regard to the vast more-than-human and ecological losses in the wake of global climate change and the currently unfolding mass extinction. Much of this scholarship builds on empirical research with people and communities that have a lived experience of losing their form of life or world as a consequence of ecological and environmental loss and destruction. An affect-oriented strand of this research investigates emotional patterns arising in the context of ecological stress, change, and devastation, such as "environmental grief", "eco-anxiety", and "climate change worry" (for an overview see Cunsolo/Ellis 2018; Ojala et al. 2021; Pihkala 2024). Another, more discourse- and practice-oriented strand of research explores forms of "mourning beyond the human" (Cunsolo/Landman 2017a), "posthumanist mourning" (Ryan 2017; Lykke 2019, 2022), and "multispecies mourning" (Chao 2023). As part of this second strand of the debate, animal studies scholars and scholars in the environmental humanities focusing on human-animal relations investigate how animals mourn (King 2014) and what it means to mourn the loss of (animal) species (Rose 2011; van Dooren 2014, 2019, 2022). Another body of knowledge, which resonates with the latter question and which I am interested in here, explores what it means to mourn the loss of vegetal life or of the ecological foundations of human and more-than-human life (e.g. Head 2016b; Cunsolo/Landman 2017b; Barnett 2022; Mihai/Thaler 2023).

These authors tie in with feminist, post- and decolonial, and Indigenous approaches to loss, grief, and mourning such as the ones sketched above for two main reasons. Firstly, the respective approaches are often rooted in a lived experience of ecological and environmental loss and unacknowledged or disavowed mourning. This applies in particular to Indigenous perspectives, which are often situated in deeply environmentally embedded naturalcultural ways of life particularly affected by environmental change and loss (see e.g. Cunsolo/Ellis 2018; Mihai/Thaler 2023; see also chapter 4.2). Secondly, focusing attention on the structural analogies between human and more-than-human forms of denied grievability, these approaches challenge the way in which "the more-than-human world often falls within the economic relations of late capitalist society and, as such, must remain ungrievable so that it can be used as a resource without consideration for its irreplaceable intrinsic value" (Mark/Di Battista 2017: 240). In contradistinction to the established resourcist approach to the more-than-human world complicit in the social-ecological devastations of the present, these authors develop a perspective on more-than-human loss built on the argument that "the ability to mourn for the loss of other species is [...] an expression of our sense of participation in and responsibility for the whole fabric of life of which we are part" (Burton-Christie 2011: 30, quoted in Cunsolo/Landman 2017a: 22). Substituting the ego- and anthropocentrism and the modernist conception of subject-object relations underlying classical understandings of grief and mourning with an attentiveness to shared and entangled lives and vulnerabilities in more-than-human communities, this body of knowledge proposes a fundamentally different understanding of the meaning and mourning of more-than-human death and loss.

A foundational and notable work in this context is Ashlee Cunsolo and Karen Landman's anthology *Mourning Nature: Hope at the Heart of Ecological Loss and Grief* (2017b). In light of the tremendous contemporary losses of species and ecosystems, which fundamentally change and challenge naturalcultural life on earth, and based on the painful awareness that humanity is implicated in these losses, both by having unleashed the destructive forces producing them and by being faced with a world that no longer sustains the dominant forms of life these forces fuel, Cunsolo and Landman ask:

> "[W]hat – politically, ethically, and theoretically – can the work of mourning do if extended beyond the human, and what might this tell us about ecological ethics, politics, and action? How does grief help us live better with others? How do we understand, think, and feel the meaning of all the ecological loss?" (2017a: 6–7)

Before I address these questions and consider how they resonate with the questions about agrobiodiversity loss and conservation raised by the art projects discussed above, the notions of grief and mourning require some explication.

There is no set definition or consistent use of these terms in the scholarly debate on grief and mourning in relation to ecological changes and losses. According to Joshua T. Barnett's synopsis of the relevant existing literature in *Mourning in the Anthropocene: Ecological Grief and Earthly Coexistence* (2022), grief often denotes a person's visceral and internal emotional response to the loss of an other or of a relationship significant to the self. Mourning, in contrast, is commonly understood as the outward expression of grief through "culturally specific practices, rituals, ceremonies, and other activities through which we adjust to the new conditions created by the loss" (Barnett 2022: 4–5). To put it in a nutshell: "If grief is a response to loss, mourning is what we do with it." (Menning 2017: 58)[30] This distinction implies a sequentiality as well as an interiority of feeling and exteriority of action that Barnett counters with an understanding of grief and mourning as "achievements", that is, "emotional, ethical,

30 The delineated differentiation between grief and mourning is rooted in the classical psychological conception of emotions as internal subjective states preceding their outward expression and social enactment. In the sociological and anthropological debate, this "'inside out' model of emotions" (Ahmed 2004: 9) is contrasted with an "'outside in' model" (ibid.) conceiving emotions as a social form and cultural practice first, which then infuses the individual body and consciousness. Criticising both models and their reification of "the very distinction between inside and outside, the individual and the social" (ibid.), Ahmed develops a relational theory of emotions and bodies as co-constitutive effects of encounters (see 2004: 5–10). This theoretical perspective, which is satisfied with neither biological (over)determinism nor social constructivism, resonates with the understanding of grief and mourning unfolded in the following.

and political capacities that must be cultivated" (2022: 5). This means, for one thing, that understanding why some losses commonly elicit grief whereas others do not requires analysing the grounds that allow for some feelings rather than others to be cultivated. To put it in Butler's terms: it demands an analysis of socio-cultural conditions of unequal grievability. For another, it implies that grief for currently ungrievable and unacknowledged losses, including more-than-human losses, *can* be cultivated. Mourning practices, which "create, nurture, or draw attention to connections" (Menning 2017: 58), can be a way to cultivate (achieve) emotional attachment where there is ontological attachment that might be known but not (yet) felt and therefore, if lost, remains unacknowledged as a loss.

The question certainly arises why feeling and grieving the vast amount of currently unfolding and impending ecological losses should be something to strive for. Let's go back to the Freudian conception of grief dominant in Western culture, according to which there are two paths in grieving: successfully working through and healing after a loss (which Freud called mourning), or being coopted by sorrow and unable to heal, thus losing oneself in one's grief over the loss of the loved other(s) (which he identified as melancholia). Despite Freud's own revision of this original theorisation, an implication of it that tenaciously persists to this day is the idea that grief is an inhibitive state of being and that "becoming 'uninhibited' is superior to remaining mired in relations with the ones we have lost" (Barnett 2022: 18). From this viewpoint, ecological grief may appear as a distraction from the struggle for social-ecological transformation and justice and the political action that needs to be taken against the destructive forces causing the current forms and dimensions of ecological loss.

In opposition to such an understanding of ecological grief and mourning as an impediment or capitulation to ecological politics, Barnett, Cunsolo and Landman, and others develop a more politicised and productive interpretation of "mourning beyond the human" (Cunsolo/Landman 2017a) or "mourning in the Anthropocene" (Barnett 2022).[31] Rather than employing a curative logic,

31 Based on a similar line of argument, Endre Dányi (2020, 2022) embraces the notion of melancholia to propose a form of politics he calls "melancholy politics". Departing from the Freudian notion of melancholy, however, Dányi develops his understanding of melancholy politics along the Hungarian term "búskomorság", which "comes close to the English word 'melancholy' but has other theoretical and political connotations" (Dányi 2020: 360). By "[r]eclaiming melancholy" (Dányi 2022: 6–27), Dányi rearticulates melancholy as a collective and resistant rather than individual and inactive sen-

they approach (ecological) grief with an affirmative perspective that highlights the ways in which grief exposes ontological relationality and interdependence (Cunsolo/Landman 2017a: 12–14; Barnett 2022: 16–19). In other words: insofar as grief makes us realise that the loss of a significant other involves the loss of a relation that was constitutive of our own life and self, grief also makes us realise our relational entanglement in the world, including our interdependent existence with the others that remain.

The work of mourning, then, involves both grieving for the lost relation(s) and strengthening other as well as fostering new life-sustaining relations that help us work through and learn to live with loss. It is important to note that "working through" a loss is not the same as "resolving" a loss in the original Freudian sense of letting go of the lost connection, restoring an integrated self by disintegrating the lost other, or striving for "full substitutability" (Butler 2004: 21; see also Cunsolo/Landman 2017a: 12). Mourning a loss and healing from it, as understood here, does not mean ceasing to feel the loss and the pain arising from it. Instead, it means integrating the loss, adapting to the pain, and finding solace, care, and endurance in the existing and potential relations that remain. Rather than restoring a prior state of stability, it means "reorient[ing] to a relatively secure and stable life in the absence of what was" (Menning 2017: 42) – "relearning our world" (Lussier 2021; see also Attig 1990) in companionship with the others that remain.[32] Mourning beyond the human, then, not only involves cultivating the grievability of more-than-human life and learning to let ourselves be affected by ecological losses. Doing so also means and, more importantly, can be a path towards, acknowledging our ontological implication in the more-than-human world and fostering life-sustaining attachments with that which remains. This, in turn, can help us work through and learn to live with the inevitable reality of ecological loss.

Following this line of argument, allowing ourselves to be affected despite the vulnerability that comes with emotional attachments, as loss painfully

---

timent that does not boil down to tristesse but can be a means of cosmopolitics. Despite the different terminological choice, Dányi's understanding of melancholy politics strongly resonates with the argument unfolded in this chapter. This illustrates the contingent use of the relevant terms.

32 Building on Haraway's notion of "becoming-with" (Haraway 2008), Lussier (2021: 145) proposes "mourning-with" as a relational and transformative practice of learning "to live and die well" with one another, emphasising that it describes "not merely an extension of 'sympathy' or 'empathy' toward the other, but the practice of rendering the other capable of mourning (without appropriating the other's grief)".

teaches us, forging caring relations by cultivating unexpected grievability and the capacity to grieve for unexpected others, is transformative both individually and collectively. As Barnett puts it: "Not only do we mourn for those we care about [...]; we in turn care about and for those whose loss we know in advance we would grieve" (2022: 20). Cultivating "anticipatory grief" (Cunsolo/Ellis 2018: 278), in other words, can be a way of relearning our world(s) and learning to foster caring kinship relations rather than extractive resource relations with the other-than-human (see chapter 4.2). Ashlee Cunsolo and Neville Ellis provide empirical evidence for such a form of grief over "anticipated future losses of place, land, species, and culture" (2018: 277). In doing so, they point out that it is an ambiguous form of grief in that it is more difficult to articulate and less likely to be publicly acknowledged than grief over something or someone already lost.

There are at least two other important ambiguities concerning anticipated and already unfolding ecological loss as well as the grief and mourning arising and practised in response to this loss. First, as I have shown throughout the course of this book, most of the ecological losses of the present are a result of centuries of resource overexploitation and environmental destruction facilitating the prosperity and progress of the modern world. Accordingly, they are more likely to be met with denial and inattention within the modern world than with grief and mourning (see e.g. Norgaard 2011). Ecological grief "in the affluent West" (Head 2016a: 81; see also 2016b), therefore, needs to be cultivated more adamantly while at the same time, its cultivation is likely to be complicated by intertwining with feelings of sorrow and guilt (see Menning 2017: 39–40).

The second additional ambiguity is that although grief is not necessarily inhibiting, it can certainly be. Against this background, anticipatory grief has a greater potential to stimulate transformative action than grief in the face of losses already experienced, not least because its trajectory is not entirely decided yet. Insofar as anticipatory grief arises from a deeply felt emotional attachment to an other, a relation, or an environment one wishes to preserve, it can become a powerful incitement to do everything in one's power to prevent the anticipated loss as well as the associated pain. Although it is certainly not possible to prevent all loss, if we feel that we can prevent a loss we would grieve we will usually aspire to prevent that particular loss. Herein lies the political potential of mourning and of extending ideas of grief and mourning beyond the confines of the current dominant order of grievability.

As can be seen, it is not the cultivation of ecological grief and the practicing of mourning for more-than-human loss as such that is the goal of a politicised understanding of ecological grief and mourning. Rather, it is the world and worlding or "reworlding" (Lussier 2021: 145) they can help cultivate (see also Butler 2004: xviii–xix). Interpreting the act of commemorating the biocultural diversity excluded from *ex situ* conservation environments such as the Svalbard Global Seed Vault, such as performed by the art projects discussed above, as a practice of mourning for more-than-human loss, opens up a political perspective on the performativity of conservation practices and rituals that harbours a transformative potential. The act of remembering biocultural connections that the art projects perform is an act of resisting the dismembering of biocultural connections in the *ex situ* world of conservation; of recollecting biocultural relations already lost as well as remaining ones that might sustain us as long as we sustain them. Rethinking conservation as commemoration, then, especially for ecologically destructive worlds, holds the potential for ecological mourning to become a political act of learning to relearn our world(s) and worldings in different and unexpected ways.

It is crucial that all this should not be understood as a 'can do' optimistic, progressivist promise of resolving the agrobiodiversity losses occurring in the wake of the unrelenting ecological destruction that comes with the current globally dominant form of more-than-human life. Where there is grief and mourning, there is pain and loss; and while mourning is a way of learning to live with loss and ease the pain that comes with it, it will never undo the loss or the pain. The mourning of ecological losses as delineated here cannot and does not serve to undo these losses, nor is it meant to ease feelings of sorrow and guilt that arise in view of perpetual destruction in order to continue with business as usual. Rather, learning to notice, acknowledge, feel, and mourn agrobiodiversity loss and other ecological losses is about learning to live with these losses and live less ruinously in the ruins of capitalist modernity so as to prevent perpetual loss. It is about a mournful welcoming of "'the end of progress' [as] not a conclusion, but a starting point" (Dányi 2022: 10) that arises from the burying of a destructive form of life that has become the greatest threat to itself. This inevitably involves the loss of ways of being in the world that are destructive but nevertheless familiar and, for many, associated with positively connotated promises of prosperity and progress. Burying and mourning destructive ways of life requires cultivating more-than-human grievability, as well as feeling rather than fearing the shared vulnerability and responsibility that comes with the interdependence of more-than-human life (see also Butler

2004: 29–30). It means fostering life-sustaining connections with that which remains; tying new ties for worlding otherwise. Then, mourning can be transformative.

Cultivating Mournful and Resistant Forms of Hope

The *Agri/Cultures.Seed-Links Exhibition* and the *Svalbard Ark* project preceding it, as documented in the film *Offerings*, perform what I conceive to be an art of mourning losses of agrobiocultural diversity. As unfolded above, both iterations of the art project are about the artworks deposited in the Arctic permafrost and the relational biocultural forms of life they represent as much as they are about the performative enactment of exhibiting, bringing to Svalbard, burying in the permafrost, and mourning these artworks. The performative art of mourning agrobiodiversity loss is an act of remembering, celebrating, and (re)connecting with the agrobiocultural diversity excluded from seed collections such as those conserved inside the Svalbard Global Seed Vault. It is an effort to remember dismembered and endangered connections between the world's seed diversity and the diversity of worlds that come with seeds, i.e. the worlds from which they emerge, in which they live, and to whose flourishing they contribute. And it is an act of collective (re)connecting with what is put in suspension, with what is around, and with one another, by sharing mourning. Unlike in the case of the Svalbard Global Seed Vault, putting something on ice, here, is not about progressivist conservationism aiming to preserve the (literal and figurative) seeds of a world on the brink of extinction so as to be able to perpetually resow it. Rather, if the art of mourning agrobiodiversity is interpreted in terms of the act of conservation it imitates, what it conserves – in the Arctic permafrost and even more so in the collective memory and repertoire of living – are naturalcultural connections (embodied by the artworks deposited in the permafrost mountain) as well as the ability to (re)connect with what has been severed (enacted through the performance that the project is), to recollect forms of life driven to the brink of extinction by the destructive forces inherent to the hegemonic form(s) of more-than-human life in the modern world.

Connected to this performative practice of mournful conservation is a hope for future recuperation. A similar form of hope is also observable in *ex situ* conservationism, though these are two different kinds of hope. In contrast to the anticipatory hope of this-worldly salvation inherent to the 'can do' optimism of *ex situ* conservation, the hope expressed in the context of the art project documented in *Offerings* remains passive. Insofar as it arises from suspending control over what one has crafted, from burying what one wishes to preserve and offering it to an unknown future, it is a form of cathartic hope for future salvation. It resembles the classical, spiritual faith that there is an afterlife beyond the confines of this world and the agency one has in it. Inherent in both the secular and the spiritual form of optimistic hope, although they differ in kind, is a lack of recognition on the part of those who experience this hope of the responsibility that comes with their entanglement in the more-than-non/human web of life, which includes their entanglement in the destructive forces continuously producing agrobio(cultural)diversity loss. Both forms of optimistic hope do not seem geared towards what Haraway (2008: 70–73, 89) calls "response-ability", that is, the ability to meaningfully respond to all this loss and destruction through one's political agency.

Importantly, as I have shown above, this lack is not inherent to *ex situ* conservation or to ecological mourning as such. It is a feature of specific performative manifestations of these practices. To give a contrasting example: In his ethnographic study of snail conservation in Hawai'i, Thom van Dooren (2022) observes a form of hope that is neither passive, nor transcendental, nor techno-utopistic. The conservationists' "mournful hope" (van Dooren 2022: 179) he describes is modest and non-utopian, but also decisive in acknowledging the increasing and often irreversible ubiquity of loss while at the same time "refusing to give up care and responsibility in the face of escalating and inevitable destruction, and to keep working toward the best relationships and possibilities that are still available to us and others" (ibid.). Van Dooren argues that this kind of "hope that is woven through with grief" (2022: 193) is the kind of "hope we need now, [...] or at least the one we get, and need to learn to live well with" (ibid.). A characteristic of this non-utopian mournful hope for ongoing life in the ruins of the past is its resistance to surrender to the reality of loss and destruction. While such a mournful, decisive, and resistant hope for and commitment to social-ecological transformation is mostly impalpable in the 2018 art project documented in *Offerings*, it resonates with the atmosphere I experienced in the *Agri/Cultures.Seed-Links Exhibition* in Svalbard in 2019. A number of the artists assembled in this second iteration of the art project exhibited works of art representing decisively non- or counter-hegemonic more-than-human ways of being in the world to which they returned after the event in Svalbard. The following two cases exemplify the resistant dimensions of their art of mourning agrobiocultural diversity loss.

The first example is the art collective *Seeds InService: A Papermaking Institute*, an ecofeminist urban gardening and papermaking art project based in Chicago

(Seeds InService 2019). The two artists, Melissa Potter and Maggie Puckett, thematically curate their gardens around ecological and feminist issues such as "heirloom seed preservation, food justice, and traditional women's healing practices" (GenØk – Centre for Biosafety 2019). By organising educational and participatory events in the gardens, they share their knowledge, experience, and practice with the public. The idea behind their gardening project, as the duo elaborates in the video interviews conducted for the Seed Cultures Archive, is to create a space in which the shared manual labour of gardening and papermaking makes it possible to experience a more-than-human process of learning and producing something in the interplay of one's own labour with that of plants, soil bacteria and microorganisms, insects, pollinators, and other more-than-human beings (see Seeds InService 2019).

The two artists describe their work as being more about the generative and potentially transformative practice and process of experiencing more-than-human relationality and interdependence through practices of gardening and papermaking than about the products of this labour. Even so, burying such products – hand-crafted paper, hand-picked seeds, and the carefully designed book *An Illuminated Feminist Seed Bank* (Potter/Puckett 2019) – for indefinite conservation in the Arctic permafrost is a mournful act of letting something go. Crucially, though, and unlike in most of the offerings of the 2018 project, this act commemorates the fruits of a processual and ongoing artistic labour that exceeds the material artwork buried in Svalbard. In this way, the hope that infuses this act and performative art of mourning what is being deposited is not a passive spiritual hope for an afterlife of that which has been buried. Rather, it is a hope that springs from the active and ongoing commitment to cultivating what could be called an "alterlife" (Murphy 2017). Alterlife describes a form of life that performatively accepts the responsibility that comes with the relationality of more-than-human life to realise one's agency in a way that cultivates less harmful more-than-human relations and futures in the harmful web of life that has come to matter in and through colonial capitalism.[33]

33 The research of M. Murphy (2017), from whom I borrow the notion of alterlife, focuses on chemical exposure, and more precisely on the often harmful relational becoming of human and more-than-human bodies with industrial chemicals. Building on and contributing to Indigenous, decolonial, and feminist STS, Murphy unfolds a decolonial feminist reading of what it means to live responsibly and persist within chemical relations. The concept of alterlife describes a figuration of life that acknowledges the relational and interdependent more-than-individual and more-than-organismic condition of life including its inescapably harmful, hurtful, and deadly manifestations. Acknowl-

A similar commitment to alterlife is discernible in the second example I wish to single out from the diversity of artworks and artists I encountered in the *Agri/Cultures.Seed-Links Exhibition*: *The Migrant Ecologies Project* founded by Singapore-based artist and educator Lucy Davis and conducted in association with the National University of Singapore Museum and the transdisciplinary master's programme Visual Cultures, Curating and Contemporary Art at Aalto University, Finland. The artwork Davis and her collaborators exhibited and deposited in Svalbard illustrates the story of a single grain of wheat found conserved in the belly of a Southeast Asian saltwater crocodile that had been dead for as long as 133 years when museum taxidermists found its gastro-intestinal tract stuffed with a large amount of wheat. *The Migrant Ecologies Project*, Davis elaborates in the video interview for the Seed Cultures Archive, traced the path of the seed from the belly of the crocodile – "this 'monster' of the tropics that has been driven out [of its habitat; FV] by colonial capital" (Davis in The Migrant Ecologies Project 2019: 3:58) – through sea routes between Singapore and India, where the wheat came from, to colonial histories of domesticating and pastoralising the tropics and supplanting Indian subsistence agriculture with export-oriented cash crop plantations. By telling the story of the more-than-human violences of colonial capitalism through the story of a kernel of wheat accidentally found in a taxidermised Southeast Asian saltwater crocodile, *The Migrant Ecologies Project* highlights that "a seed [...] isn't necessarily a benevolent representative of a benevolent history" (Davis in The Migrant Ecologies Project 2019: 6:13). It shows that more-than-human relations and the ideas they are cultivated to facilitate, such as prosperity and progress, are historically situated; that more-than-human relations, while generative for some, can be violent and devastating for others.

Through the artwork exhibited and deposited in Svalbard, *The Migrant Ecologies Project* illustrates that there are different stories and histories to tell with the stories of seeds, and that remembering and conserving seeds

---

edging that life is "entangled within community, ecological, colonial, racial, gendered, military, and infrastructural histories that have profoundly shaped the susceptibilities and potentials of future life" (Murphy 2017: 497), alterlife embraces this exposure to alteration. In a decisive openness to alteration, Murphy argues, alterlife recognises and seizes "the potential to become something else, to defend and persist, to recompose [more-than-human] relations [...], to become alter-wise in the aftermath" (2017: 500). I admit to bending the authorial intent to a certain degree here by transposing the category into a very different context. The reference is an inspiration more than an analytical argument in the narrower sense.

and their stories is not an innocent act but entangled in political choices. Remember Haraway: "it matters what stories we tell to tell other stories with" (2016: 12). The choice to bury and conserve the violent and destructive more-than-human (hi)stories of colonial capitalism embodied in the crocodile, and the artwork representing it, illustrate a form and performance of commemorating and mourning violent more-than-human relations that is resolutely resistant. What distinguishes such "resistant mourning" (Cunsolo/Landman 2017a: 14–15), as Cunsolo and Landman describe it, is that it is not future-oriented but stays with the past and present by working to expose and protest the injustices and oppressive structures that cause loss in the first place.[34] Resistant mourning decisively seeks more than finding hope in learning to "live well with" loss. It is a form of mourning committed to alterlife. It strives not for consolation, compensation, or catharsis, but for the recognition of losses as losses and for accountability and response-ability vis-à-vis the destructive forces that turn forms of life into forms of loss. Resistant mourning thus also makes remembering these forces part of the commemoration, as a way of ensuring that the loss and devastation they unleash will not fall into oblivion.

In light of the devastating reality and ever more existential threat of global agrobiodiversity loss, the Svalbard Global Seed Vault and the genebanks around the world it reinsures have created a safety net that is able to secure a large and constantly growing diversity of vegetal lifeforms from extinction. While this work is invaluable in the given reality of ongoing loss and destruction, it fails to recognise the equally endangered and disappearing diversity of forms of life that (agro)biodiversity is rooted in. Aside from the fact that cultural de-diversification is a manifestation of the imperialist and violent one-world worldism of colonial capitalism and worthy of critique for that reason alone, much of the biological diversity persevering and continuing to evolve *in situ* today exists in and depends on spaces cultivated by non-modernist agri-food-systems. As I showed through the discussion of global interdependence in chapter 4.2, the *ex situ* world of conservation is well aware of this. However, instead of accounting for the structural entanglement of modern agriculture in the losses with which it is confronted, and adapting the system to the naturalcultural realities it faces rather than the other way around, for example

34 Cunsolo and Landman adopt the notion of resistant mourning from the work of Clifton Spargo (2004) and Patricia Rae (2007), who introduced it to describe a form of AIDS activism that has sought to ensure recognition of AIDS-related losses by convening public mourning ceremonies to resist stigmatisation.

by broadening the scope of agrobiodiversity conservation, the *ex situ* world promises to conserve modern agriculture by technologically circumventing its systemic destructiveness. Inasmuch as the conservation of endangered agrobiodiversity does not entail the conservation of agrobiocultural diversity and forestalls social-ecological transformation in the centres of modern agriculture towards cultivating more diversity and de-cultivating the de-diversifying forces that endanger and destroy agrobiodiversity, it ultimately conserves not only agrobiodiversity but the roots of agrobiodiversity loss as well.

To understand the tenacity of this vicious circle of loss and conservation and learn how to break out of it, it is important to recognise that the *ex situ* world's techno-optimistic approach to agrobiodiversity loss and conservation, just like the passive hope for an afterlife as its spiritual equivalent and game-over-apocalypticism as its defeatist counterpart, is a characteristic affective response to "the horrors of the Anthropocene and the Capitalocene" (Haraway 2016: 3). The problem with these affects, as I argued in chapter 4.2, is their "abstract futurism" (Haraway 2016: 4) and the concomitant lack of attention to the troubles, responsibilities, and response-abilities of the "complex worlding [...] game of living and dying well together" (Haraway 2016: 29) in the present. Crucially, this does not mean, either for Haraway or for me, condemning technoscientific approaches to agrobiodiversity loss and the *ex situ* world of conservation altogether. My critique targets the adherence of the technoscientific project of *ex situ* conservation to the world and the destructive worlding practices it has grown out of.

As I have shown throughout this chapter, recognising these destructive entanglements of the *ex situ* world of conservation, burying them, and mourning the familiar ways of life that will be lost in this process can be a way of learning to feel rather than fear agrobiodiversity loss and of learning to live with it. It can be a performative path towards cultivating less destructive forms of shared more-than-human life. The practice of conserving seeds *ex situ* is not a techno-futuristic project per se, nor is it determined by its currently hegemonic context. The world of conservation comprises other worlds and modes of worlding that can – and do already, as in the case of the collaboration of *ex situ* conservationists with the Potato Park (see chapter 4.2) – inspire ways of doing conservation and dealing with loss differently, beyond the *ex situ* world(ing).

Inspired by the *Agri/Cultures.Seed-Links Exhibition* and the *Svalbard Ark* project, which the Seed Vault has assembled on the margins of the *ex situ* world of conservation, I have proposed to rethink agrobiodiversity conservation through the performative art of mourning agrobiodiversity loss. Mourning is a practice that allows for processing the experience of loss and adapting to an altered reality while also inspiring hope, transformative action, and resistance to persevering destructive forces. Thus understood, a collective practice of mourning the vast ecological and social-ecological losses of the present and future can be a performative and transformative path towards learning to live with them rather than enduring despite them. It can incite a hope that is mournful and resistant and hence comes with a commitment to ongoingness not in spite of the social-ecological troubles that come with agrobiodiversity loss but by way of staying with these troubles.

6. Coda

The Svalbard Global Seed Vault, as the people and institutions behind it envisioned, designed, realised, use, and narrate it, is a "simple and straightforward" (Fowler 2008: 501) response to an existential social-ecological problem: the endangerment and loss of plant genetic diversity for food and agriculture both in farmers' fields (*in situ*) and in genebanks (*ex situ*). In *Seeds on Ice*, Cary Fowler describes the Seed Vault's "quiet rescue mission" (2016: 17) as follows: "It is about hope and it is about commitment – about what can be done if countries come together, shed suspicion and cynicism, and work cooperatively to accomplish something significant, long lasting, and worthy of who we are and wish to be." (Fowler 2016: 18) The sociological inquiry into the Svalbard Global Seed Vault and agrobiodiversity conservation I have developed throughout this book has ultimately put a question mark over this account by asking: What exactly is it that the Svalbard Global Seed Vault is committed to? Who is the "we" whose hopes and commitments it brings together? And what precisely does this "we" "wish to be" and "accomplish" through the Seed Vault? Put simply, I have discussed what kind of a solution the Seed Vault effectively offers to the problem of global agrobiodiversity loss, or what exactly it ultimately acts as a solution and conservation technology for.

To do so, I have focused analytical attention on the Seed Vault not primarily as a facility and technoscientific instrument for long-term seed conservation – although that is certainly what it is. As a backup storage facility for the crop diversity held in seed collections all over the world, however, the Seed Vault is also a nodal point of a diversity of globally distributed agrobiodiversity conservation efforts, all situated in their own local worlds and histories. It is precisely this functioning of the Seed Vault as not only a technoscientific apparatus for crop diversity conservation but also an arena assembling a diversity of worlds and modes of world-making that I have put at the centre of my inquiry. Through this lens, I was able to show that the Seed Vault and the kind of agrobiodiversity conservation it has been and might yet be capable of achieving come to matter in the interplay of a diversity of actors, practices, technologies, discourses, rationalities, stories, histories, imaginaries, hopes, and more. I want to conclude this book by reflecting on the implications of my study, firstly for the question of how to understand and approach agrobiodiversity loss and conservation and secondly for sociological engagement with contemporary social-ecological changes and challenges. With regard to the latter, what follows is not a conclusion so much as a reopening of the case of agrobiodiversity loss and conservation – an invitation to reckon with how certain ways of knowing and worlding the world are implicated in the losses 'we' seek to avert or adapt to and to reconsider how we know and world our world(s).

6.1 Rethinking Agrobiodiversity Loss and Conservation Through the Svalbard Global Seed Vault

To understand the complexity of the global social-ecological threat of agrobiodiversity loss and the ambiguities of the *ex situ* approach to agrobiodiversity conservation, it is crucial to historicise both agrobiodiversity loss and conservation. I began this book by situating the history of international agrobiodiversity conservation efforts from the mid-twentieth century onwards, which eventually led to the creation of the Svalbard Global Seed Vault, within the historical context of agricultural 'modernisation' (see chapter 1). This enabled me to show that the large-scale loss of agricultural biodiversity facing the world as a global social-ecological problem today is as much a *challenge for* as it is a *product of* modern agriculture and the modern world more generally. Through processes of crop homogenisation and the introduction of agro-chemicals into farming all around the world, which promised to make global agricultural production more productive and efficient, the modernisation and globalisation of agriculture in the second half of the twentieth century provoked the worldwide decline in agrobiodiversity that became the global ecological problem of "genetic erosion" (Frankel et al. [1969] 1970; Khoury et al. 2022; see chapter 1.2).

The rise of *ex situ* conservation as the key international strategy for saving the world's plant genetic diversity from severe erosion is a manifestation of the international desire to maintain rather than challenge the modes of production underlying the loss of agricultural biodiversity while mitigating its existential effects. Similarly, the creation of the Svalbard Global Seed Vault as a safety backup for the global system of genebanks is a response to the vulnerability of the latter, in much the same way as genebanks are a response to the vulnerabilities of modern agricultural production. The irony of this approach to saving agrobiodiversity from loss is the following: conserving a plant genetic diversity *ex situ* that is not being cultivated *in situ*, in order to maintain a supposedly more efficient global monocultural agriculture, ultimately conserves the resourcist nature relations of modern agricultural production – and hence the very conditions for the (ongoing) loss, or rather destruction of the world's agrobiodiversity.

It seems reasonable, then, to raise the question whether *ex situ* conservation is an appropriate response to agrobiodiversity loss at all. Some critics in the non-governmental and civil society sector argue, along these lines, that the Svalbard Global Seed Vault evokes "a false sense of security" (GRAIN 2008: 1) in a world that continues to destroy agrobiodiversity *in situ* and that a more appropriate response to large-scale agrobiodiversity loss than conserving plant genetic diversity *ex situ* is cultivating agrobiodiversity *in situ* (see chapter 3.1). While there is some force to this problematisation, the dichotomic juxtaposition of *ex situ* and *in situ* conservation approaches overlooks the ambiguous ways in which the Seed Vault and *ex situ* conservation, like other technologies and practices of agrobiodiversity conservation, are situated and entangled in modes of knowing, being in, and enacting the world that are not reducible to either of these conservation approaches. Paying attention to the more-than-human relations, knowledges, and histories that divergent practices, technologies, and rationalities of agrobiodiversity loss and conservation are embedded in and enact allows for a more complex understanding of the more-than-human worlds and futures they give rise to.

The question that has particularly interested me in this book, therefore, is not which approach or technology is able to preserve agrobiodiversity more efficiently or sustainably, but which world(s) and modes of worlding conservation technologies and approaches are engaged in sustaining. Against this background, I have not centred analytical attention on the agrobiodiversity conservation efforts *of* the Seed Vault, but investigated the diversity of agrobiodiversity conservation efforts that become discernible *through* the Seed Vault. This has enabled me to show that, while the Seed Vault is an outgrowth of a world built on a "resourcist cosmovision" (Fenzi/Bonneuil 2016: 78) that sees and enacts the natural/ised world as a set of extractible resources, it is neither determined by nor necessarily restricted to reproducing this world and mode of worlding. Rather, as an arena assembling not only a *diversity of seeds* but also a *diversity of worlds*, the Seed Vault makes the question of what world/s and worlding/s it serves to sustain a matter of collective negotiation. A pertinent example of this is provided by collaborations with Indigenous depositors such as the Andean Potato Park, which not only bring seeds into the *ex situ* world but also confront the latter with fundamentally different modes of knowing, being in, and enacting the world (see chapter 4.2).

Through the contraposition of different worlds and modes of worlding the Seed Vault assembles, I have further raised the question of how to think about and address the issue of agrobiodiversity *loss*. In the *ex situ* world of conservation, loss of agrobiodiversity is tantamount to the loss of plant genetic resources required for adapting crops and agricultural production to changing ecological conditions. Insofar as more diversity holds more potential for successful adaptation, conserving plant genetic resources *ex situ* appears to be the obvious choice for managing *in situ* agrobiodiversity loss (see chapter 4.1). In this frame of thought, the prospect of having as many resources as possible to assist adaptation to all kinds of future environmental changes and challenges fosters faith in a plentiful rather than threatful future (see chapter 5.1). Underlying this understanding of and relation to loss is a resourcist understanding of and relation to the natural/ised world.

That the latter is not a universal ontology so much as a historically specific, situated way of knowing, being in, and enacting the world becomes discernible, again, in the example of the Potato Park. There, losing biological diversity, for one thing, means losing nonhuman community or family members. For another, it entails a loss of cultural identity and potentially one's world. Thus understood, loss is not only a problem to be managed, e.g. by means of technoscientific adaptation, but also an affective experience (both personal and collective) that may come with feelings and practices of grief and mourning. Through the examination of an art project in Svalbard commemorating the biocultural diversity excluded from the conservation efforts of the Svalbard Global Seed Vault, I have discussed what it means to rethink conservation in the context of an affective acknowledging of agrobiodiversity loss as an inevitable past, present, and future reality (see chapter 5.2). Aside from underlining the vital importance of conserving a diversity of worlds rather than merely a diversity of resources for a world cultivating agrobiodiversity loss, the example of the art project enabled me to explore the potential of performative practices of grief and mourning to counter the techno-optimism of the *ex situ* world of conservation with a "mournful hope" (van Dooren 2022: 179) that it may be possible to cultivate less destructive worlds.

6.2 Collecting Seeds of a Critical More-Than-Human Sociology For (or Against) the Anthropocene

My inquiry into the world of agrobiodiversity conservation through the Svalbard Global Seed Vault illustrates and performs the kind of sociological analysis of contemporary technoscientific and other world-making efforts I am convinced the current social-ecological condition urgently calls for. The present is a time of intersecting social-ecological catastrophes and transformations that are shaking the foundations and certainties of the modern world. This has become paradigmatically evident in the proposal scientists have advanced at the beginning of the twenty-first century, the idea that humanity's massive interferences in the earth's ecosphere have heralded the start of a new geological era after the Holocene, the Anthropocene (Crutzen/Stoermer 2000; Crutzen 2002).

The incessant multi- and transdisciplinary debate that has unfolded around the idea of the Anthropocene since then has drawn the attention of scholars across all academic disciplines as well as the non-academic public to a reality that feminist and decolonial theorists, Indigenous thinkers, and with them a number of critical anthropologists and sociologists have been pointing out for a long time: that the human, the cultural, the social, the technological, etcetera, on the one hand and nature, on the other hand, are not the ontologically distinct spheres that dominant Western thinking holds them to be. The Anthropocene debate has given rise to a quantitatively – if not qualitatively – unprecedented awareness of the fact that nature is not an ahistorically given, passive, non-agential surface, environment, and endlessly exploitable reservoir of resources with no value or purpose of its own beyond serving human 'development', but rather a dynamic, resourceful as well as recalcitrant, material reality deeply entangled in and co-emerging with societal formations that are always already and inevitably more-than-human.

The Anthropocenic awakening of the modern world (note: not *the* world) has fundamental implications for both the natural sciences and the social sciences and humanities. To this day, scientific knowledge production is to a large degree informed by the classical separation of the humanities and social sciences, on the one hand, as the academic authorities over largely human-centred ideas of culture, social life, actions, collectives, discourses, structures, power relations, etcetera, and the natural sciences, on the other hand, as the academic domain in charge of the nonhuman and natural/ised world. However, there are various bodies of knowledge in different academic disciplines as well as inter- and transdisciplinary research fields that challenge this separation, highlighting the many entanglements of the supposedly distinct spheres instead. In this book, I have used the example of biodiversity loss and conservation to show how feminist and decolonial theory as well as debates in anthropology, STS, and political ecology can productively enrich a critical, post-anthropocentric sociology of social-ecological devastations and transformations.

It has been almost half a century since the US sociologists William R. Catton and Riley E. Dunlap (1978) first criticised the "Human Exceptionalism Paradigm" of sociological thought and made the case for a "New Environmental Paradigm" that should focus attention on the interdependencies between the social and the ecosphere. Their work was foundational for the emergence of environmental sociology as a subdiscipline of US sociology in the 1970s and also informed the emergence of environmental sociology in the European context in the 1980s. However, whereas the new subdiscipline of environmental sociology turned environmental attitudes, discourses, policies, and activism into a subject of sociological analysis, the paradigm shift Catton and Dunlap had advocated failed to materialise (including in their own research). Although environmental sociology has evolved into a thriving and multifaceted field of research in the past fifty years, it has largely remained a "sociology of environmental issues" (Dunlap/Catton 1979: 244).

As a result, the classical dichotomic understanding of nature and society as two different, albeit interacting spheres has largely remained unchallenged. Environmental sociologists have certainly developed or embraced approaches and concepts that theorise the interrelations and interdependencies of natural and societal realities such as social ecology, societal nature-relations, and environmental justice. However, more radically relational approaches that emphasise the ontological inseparability of the human from nature as well as the constructionist character and historical specificity of their separation, as have been developed for more than three decades in ecofeminism (e.g. Plumwood 1993; Sturgeon 1997), feminist materialisms (e.g. Haraway 1989, 1991; Barad 2003, 2007), anthropology (e.g. Viveiros de Castro 1998, 2015; Descola 1994, 2014), and decolonial political ecology (e.g. Escobar 1999; Alimonda 2011, 2019), remain rare.[1]

---

1 Notable recent exceptions can be found in Leinius (2021) and Sempértegui (2021). It should also be noted that in the recent past, interest in the mentioned fields of research has grown in (environmental) sociology. Movements of thought such as the new

Corresponding to this epistemological profile, environmental sociology empirically fosters interdisciplinary collaborations that bridge the disciplinary separation of social and natural scientific expertise but leave its epistemological foundations intact. The social-ecological situation that the Anthropocene debate draws attention to, however, challenges the ontological segregation of the social and the natural in a way that requires not only bridging epistemic gaps but querying their epistemological foundations. What historians Christophe Bonneuil and Jean-Baptiste Fressoz have called "the shock of the Anthropocene" (Bonneuil/Fressoz 2017) is a fundamental challenge not only to modern ways of *being in* the world, but also and concomitantly to the hegemonic modern way of *knowing* the world. It has disrupted the modern certainty that the human is essentially and ontologically distinct from nature, as well as the assumption that nature is a passive surface and a reservoir of resources available for humans, societies, or cultures to occupy and use. The Anthropocene experience makes tangible, including for many of those who have heretofore managed to escape this reality, what feminist, post- and decolonial, Indigenous, and other critical voices including the ones cited above and many more have been emphasising for a long time: that the dualistic ontological divide between the Human and Nature is not a universal truth, but the product of a historically specific, geo- and biopolitically embedded, and all but uncontested epistemology.

This onto-epistemology has historically served as a naturalising justification for centuries of appropriating and exploiting land, natural/ised resources, and labour (both nonhuman and human) as well as other ecologically destructive modes of being in, subjugating, and exploiting worlds for the prospering of the modern world. That world was built on the enactment of the idea that the "domination [of nature] is simply 'natural', flowing from nature itself and the nature(s) of things" (Plumwood 1993: 4). Insofar as the current catastrophic social-ecological condition is a material result of this mode of knowing and enacting the world, it compromises the modern world and world-making project ecologically as much as onto-epistemologically. To put it in the words of the

materialisms and the ontological turn in social theory in particular have engendered a sociological reckoning with radically relational understandings of nature and society or culture (see e.g. Pellizzoni 2016; Folkers 2020; Engert/Schürkmann 2021; Lemke 2021; Schürkmann 2021; Barla 2023; Hoppe 2024; Silberzahn 2024; see also Kropp/Sonnberger 2025: 60–76).

Colombian-American decolonial anthropologist Arturo Escobar, the Anthropocene – or rather the social-ecological predicament that this term so insufficiently encapsulates – is an "environmental-cum-ontological conflict" (Escobar 2020: xxxi).

Thus understood, the social-ecological devastations and challenges of the present urge academic knowledge production to change the *terms* rather than merely the *content* of the conversation about the ecological crises facing the world (Mignolo 2011: 122–123; 2018: 143–144; see also von Verschuer/Hartmann/Barla 2023). In order to achieve a comprehensive sociological understanding of the current social-ecological condition, it is essential to attend to how ways of knowing and being in the world are implicated in destructive modes of social-ecological world-making. This applies particularly to a sociology pursuing both an empirical-analytical and a critical-transformative approach, as much contemporary environmental sociology (including my own) does. In order to do so, an inter- and transdisciplinary opening towards other bodies of knowledge besides the natural sciences is crucial. It is not only the natural and the social sciences that can enrich one another in their engagement with the social-ecological conditions, transformations, and challenges of the past, present, and future. Thinking beyond disciplinary boundaries within the domain of the social sciences and humanities, thinking with anthropology, political ecology, feminist materialisms and technoscience studies, decolonial thought, and Indigenous scholarship holds just as much potential for environmental sociology and sociology more generally for mutual inspiration and learning.

In the world-historical timespace that concepts such as the Anthropocene, the Capitalocene or the Chthulucene circumscribe, in which it appears to be the world itself that is at stake, it is crucial to carefully scrutinise what world we are talking about. Something that is emphasised in all bodies of knowledge I have woven into my exploration of the world/ings of agrobiodiversity conservation is that "the world" is not an ahistorically given entity to be known and used. Instead, as I have shown throughout this book, what the world is comes to matter – for better or for worse – in emergent and indeterminate ways and in (sometimes unexpected) more-than-human relations and collaborations. Acknowledging the diversity of ways of knowing, being in, and enacting the world – the diversity of modes of worlding – and creating arenas where world(ing)s meet holds the potential, to borrow the words of my interview partner Alejandro Argumedo once more, of cultivating relations of "mutually beneficial cross-fertilisation of ideas and experiences" (see chapter 4.2). It is, in other words, an opportunity for and crucial step towards cultivating less destructive worlds.

Acknowledgments

> To be one is always to *become with* many.
> (Haraway 2008: 4, emph. in orig.)

For a seed to grow into a plant that blossoms and eventually produces new seeds, a lot of collaborative labour and labourers are essential. Light, water, pollinators, fertile soil, bacteria, and other living organisms are indispensable to the vegetal life cycle. Similarly, this book has grown out of a deeply relational process of learning, questioning, searching, assembling, weaving, doubting, unravelling, rethinking, and reweaving. It would not have become what it is without the many others who have supported and nourished this process – especially since it came to life in a time of many troubles shaking both the world at large and my very own personal world. I am deeply grateful to all the people who accompanied, inspired, and encouraged me on this journey, many of whom inhabit these pages without the citational credit others have received.

This book is a revised version of my doctoral thesis, which I have submitted to Goethe University Frankfurt in June 2024. First, I want to thank my supervisors, Thomas Lemke and Susanne Schultz, for their generous support, friendly guidance, and stimulating challenges. My sincerest appreciation also goes to my interview partners and other informants who have kindly agreed to share their knowledge with me. These encounters have enriched my thinking in more ways than the passages quoted throughout this book can display. I thank Viona Hartmann and Paula Stiegler for their support in transcribing the interviews. My ethnographic research trips were made possible by the kind support of the Vereinigung der Freunde und Förderer der Goethe-Universität (Association of Friends and Sponsors of Goethe University), the GRADE Center Social Sciences, GRADE Center Gender, and the Department of Social Sciences at Goethe University.

I could not have written this book, nor would the process have been so joyful without the relentless encouragement, unparalleled support, and great friendship of Josef Barla, Katharina Hoppe, Vicky Kluzik, Christoph Panzer, and Lucas Schucht, who have provided crucial comments on several of the chapters and offered invaluable help in figuring out the final structure. I am further deeply grateful to other colleagues and friends who have encouraged and supported me in the past years through (mutual) proofreading and collaborative analysis of empirical material, in-depth discussions and spontaneous conversations, joint events and collective writing sessions: Sophie Bauer, Julia Böcker, Marleen Boschen, Veit Braun, Franziska Dahlmeier, Endre Dányi, Karin Dick, Constanze Erhard, Andreas Folkers, Louka Maju Goetzke, Julia Hartleb, Viona Hartmann, Anastassija Kostan, Anna-Katharina Laboissière, Sara Lafuente-Funes, Ruzana Liburkina, Ruth Manstetten, Carolin Mezes, Pierre Louis du Plessis, Nils Richterich, Markus Rudolfi, Jonas Rüppel, Eva Sänger, Luki Schmitz, Laura Schnieder, Aline Theobald, Annika Troitzsch, Peter Wehling, Tobias Wagner, Leon Wolff, Franziska Zirker, and Ira Zöller.

I would also like to express my gratitude to Ursula Münster for inviting me to join the Oslo School of Environmental Humanities as a visiting researcher in the spring of 2023 and affording me the chance of many inspiring encounters and conversations. I thank the students who have participated in the seminars I had the pleasure to teach at Goethe University over the past years and the participants of the Biotechnologies, Nature and Society Colloquium for inspiring discussions and collaborative learning. I would also like to thank the colleagues I worked with in the Women's and Gender Equality Council of the Faculty of Social Sciences at Goethe University between 2019 and 2021, who taught me a lot about the power of working in an atmosphere of feminist solidarity. A special thanks goes to Renate Uhrig and Angelika Boese for administrative and technical support always infused with care and compassion. I thank Julian Koptisch and Moritz Dapper for their editorial support and Gerard Holden for copy-editing the final manuscript.

My heartfelt thanks also go to my friends and family who shared the many challenges as well as the joys of the past years with me. I want to express my deepest appreciation to my partner Wolfgang, whose unrelenting care, confidence, enthusiasm, humour, and love, more than anything, have carried me through this journey with all its ups and downs. Finally, no words can express the gratitude I feel for my beloved mother Katja, whose unbearable loss I had to survive amidst all of this. You taught and never stopped encouraging me to follow my curiosity, trust my gut, speak my mind, and maintain my wit even in the most disheartening of times. You will forever be my greatest source of inspiration.

Index of Abbreviations

AHTEG	Ad Hoc Technical Expert Group
BSF	Benefit Sharing Fund
CBD	Convention on Biological Diversity
CGIAR	Consultative Group on International Agricultural Research
CGRFA	Commission on Genetic Resources for Food and Agriculture (as of 1995; formerly CPGR)
CIAT	International Center for Tropical Agriculture (Abbreviation for Spanish: Centro Internacional de Agricultura Tropical)
CIMMYT	International Maize and Wheat Improvement Center (Abbreviation for Spanish: Centro Internacional de Mejoramiento de Maíz y Trigo)
CPGR	Commission on Plant Genetic Resources (renamed CGRFA in 1995)
CSO	Civil Society Organization
ETC Group	Action Group on Erosion, Technology and Concentration (as of 2001; formerly RAFI)
FAO	Food and Agriculture Organization of the United Nations
GPA	Global Plan of Action for the Conservation and Sustainable Utilization of Plant Genetic Resources for Food and Agriculture
IAP	International Advisory Panel
IARC	International Agricultural Research Centre
IBP	International Biological Programme
IBPGR	International Board for Plant Genetic Resources (renamed IPGRI in 1996)
ICARDA	International Center for Agricultural Research in the Dry Areas
IPGRI	International Plant Genetic Resources Institute (as of 1996; formerly IBPGR; renamed Bioversity International in 2006)

IPK	Leibniz Institute of Plant Genetics and Crop Plant Research (Abbreviation for German: Leibniz-Institut für Pflanzengenetik und Kulturpflanzenforschung)
ITPGRFA	International Treaty on Plant Genetic Resources for Food and Agriculture (short: Seed Treaty)
IU / IUPGR	International Undertaking on Plant Genetic Resources
MLS	Multilateral System
MSBP	Millennium Seedbank Partnership
NGO	Non-Governmental Organisation
PGR	Plant Genetic Resources
PGRFA	Plant Genetic Resources for Food and Agriculture
RAFI	Rural Advancement Foundation International (renamed ETC Group in 2001)
SDGs	Sustainable Development Goals
SGSV	Svalbard Global Seed Vault
SIS	Svalbard International Seedbank
SMTA	Standard Material Transfer Agreement
STS	Science and Technology Studies
UNEP	United Nations Environment Programme
UPOV	International Union for the Protection of New Varieties of Plants (Abbreviation for French: Union Internationale pour la Protection des Obtentions Végétales)

References

Adamson, Joni/Monani, Salma (Eds.) (2017): Ecocriticism and Indigenous Studies. Conversations from Earth to Cosmos, New York, London: Routledge.

Ahmed, Sara (2004): The Cultural Politics of Emotion, Edinburgh: Edinburgh University Press.

Alimonda, Héctor (2019): »The Coloniality of Nature. An Approach to Latin American Political Ecology«, in: Alternautas 6 (1), pp. 102–142.

Alimonda, Héctor (Ed.) (2011): La Naturaleza Colonizada. Ecología Política y Minería en América Latina [Colonised Nature. Political Ecology and Mining in Latin America], Buenos Aires: CLACSO.

Alpsancar, Suzana (2016): »Plants as Digital Things. The Global Circulation of Future Breeding Options and their Storage in Gene Banks«, in: Italian Journal of Science and Technology Studies 7 (1), pp. 45–66.

Anderl, Felix (2024): Epistemologies of Land, Blue Ridge Summit: Rowman & Littlefield.

Andersen, Regine (2003): »FAO and the Management of Plant Genetic Resources«, in: Olav S. Stokke/Øystein B. Thommessen (Eds.), Yearbook of International Co-Operation on Environment and Development, 2003/2004, London: Earthscan, pp. 43–53.

Andersen, Regine (2008): Governing Agrobiodiversity. Plant Genetics and Developing Countries, Burlington: Ashgate.

Anderson, Ben (2009): »Affective Atmospheres«, in: Emotion, Space and Society 2 (2), pp. 77–81.

Anderson, Ben/Grove, Kevin/Rickards, Lauren et al. (2020): »Slow Emergencies. Temporality and the Racialized Biopolitics of Emergency Governance«, in: Progress in Human Geography 44 (4), pp. 621–639.

Arctic Monitoring and Assessment Programme (2012): Arctic Climate Issues 2011. Changes in Arctic Snow, Water, Ice and Permafrost. SWIPA 2011 Overview Report, Oslo.

Argumedo, Alejandro (2008): »The Potato Park, Peru. Conserving Agrobiodiversity in an Andean Indigenous Biocultural Heritage Area«, in: Thora Amend/Jessica Brown/Ashish Kothari et al. (Eds.), Protected Landscapes and Agrobiodiversity Values, Heidelberg: Kasparek, pp. 45–58.

Argumedo, Alejandro/Song, Yiching/Khoury, Colin K. et al. (2020): »Support Indigenous Food System Biocultural Diversity«, in: The Lancet Planetary Health 4 (12), p. e554.

Argumedo, Alejandro/Song, Yiching/Khoury, Colin K. et al. (2021): »Biocultural Diversity for Food System Transformation Under Global Environmental Change«, in: Frontiers in Sustainable Food Systems 5, pp. 1–9.

Arlov, Thor B. ([1989] 1994): A short History of Svalbard, Oslo: Norsk Polarinstitutt.

Arlov, Thor B. (2005): »The Discovery and Early Exploitation of Svalbard. Some Historiographical Notes«, in: Acta Borealia 22 (1), pp. 3–19.

Åsberg, Cecilia/Lykke, Nina (2010): »Feminist Technoscience Studies«, in: European Journal of Women's Studies 17 (4), pp. 299–305.

Asdal, Åsmund (2025): »The Svalbard Global Seed Vault – Conserving Plant Genetic Resources for European and Global Food Security«, in: Genetic Resources (S2), pp. 49–57.

Asdal, Åsmund/Brodal, Guro/Solberg, Svein Ø. et al. (2019): »Seed Longevity and Survival of Seed Borne Diseases after 30 Years Conservation in Permafrost. Report from the 100 Year Storage Experiment«, in: NordGen Publication Series 2019: 04, Alnarp, Sweden, http://norden.diva-portal.org/smash/get/diva2:1370439/FULLTEXT02.pdf from June 10, 2025.

Asociación ANDES (n.d.): Potato Park – PISAQ, https://andes.org.pe/potatopark-pisac/ from June 10, 2025.

Asociación ANDES/Black Maria (2016): UYWAY (The Seed), https://vimeo.com/200658165 from June 11, 2024.

Asociación ANDES/Potato Park (2015): Biocultural Heritage Innovations in the Potato Park. SIFOR Qualitative Baseline Study, Peru, London: International Institute for Environment and Development (IIED), https://www.iied.org/sites/default/files/pdfs/migrate/G03917.pdf from June 10, 2025.

Attig, Thomas (1990): »Relearning the World. On the Phenomenology of Grieving«, in: Journal of the British Society for Phenomenology 21 (1), pp. 53–66.

Austin, J. L. (1975): How to do Things with Words, Cambridge, MA: Harvard University Press.

Barad, Karen (2003): »Posthumanist Performativity. Toward an Understanding of How Matter Comes to Matter«, in: Signs: Journal of Women in Culture and Society 28 (3), pp. 801–831.

Barad, Karen (2007): Meeting the Universe Halfway. Quantum Physics and the Entanglement of Matter and Meaning, Durham: Duke University Press.

Baranski, Marci R. (2022): The Globalization of Wheat. A Critical History of the Green Revolution, Pittsburgh: University of Pittsburgh Press.

Bargh, Maria (2021): »Diverse Indigenous Environmental Identities. Māori Resource Management Innovations«, in: Brendan Hokowhitu/Aileen Moreton-Robinson/Linda T. Smith et al. (Eds.), Routledge Handbook of Critical Indigenous Studies, Abingdon, New York: Routledge, pp. 420–430.

Barla, Josef/von Verschuer, Franziska (2022): »Anthropocene«, in: Matter: Journal of New Materialist Research 2 (3), pp. 137–143.

Barnett, Joshua (2022): Mourning in the Anthropocene. Ecological Grief and Earthly Coexistence, East Lansing: Michigan State University Press.

Barnosky, Anthony D./Matzke, Nicholas/Tomiya, Susumu et al. (2011): »Has the Earth's Sixth Mass Extinction Already Arrived?«, in: Nature 471 (7336), pp. 51–57.

Battistoni, Alyssa (2017): »Bringing in the Work of Nature. From Natural Capital to Hybrid Labor«, in: Political Theory 45 (1), pp. 5–31.

Beck, Ulrich (1992): Risk Society. Towards a New Modernity, London: SAGE.

Beck, Ulrich (1999): World Risk Society, Cambridge, UK: Polity Press.

Beck, Ulrich (2004): »The Truth of Others. A Cosmopolitan Approach«. Symposium: Talking Peace with Gods, Part 1, in: Common Knowledge 10 (3), pp. 430–449.

Beck, Ulrich/Wehling, Peter (2012): »The Politics of Non-Knowing. An Emerging Area of Social and Political Conflict in Reflexive Modernity«, in: Fernando D. Rubio/Patrick Baert (Eds.), Politics of Knowledge, Abingdon: Routledge, pp. 33–57.

Bell, Catherine M. (1992): Ritual Theory, Ritual Practice, New York: Oxford University Press.

Bellah, Robert N. (2003): »The Ritual Roots of Society and Culture«, in: Michele Dillon (Ed.), Handbook of the Sociology of Religion, Cambridge, UK: Cambridge University Press, pp. 31–44.

Belliger, Andréa/Krieger, David J. (2008): »Einführung« [»Introduction«], in: Ritualtheorien. Ein einführendes Handbuch [Ritual Theory. An Introductory Handbook], Wiesbaden: VS Verlag für Sozialwissenschaften, pp. 7–34.

Benner, Susanne/Lax, Gregor/Crutzen, Paul J. et al. (Eds.) (2021): Paul J. Crutzen and the Anthropocene. A New Epoch in Earth's History, Cham: Springer.

Bhambra, Gurminder K. (2014): »Postcolonial Entanglements«, in: Postcolonial Studies 17 (4), pp. 418–421.

Blaser, Mario (2016): »Is Another Cosmopolitics Possible?«, in: Cultural Anthropology 31 (4), pp. 545–570.

Blumenberg, Hans ([1960] 2009): Paradigmen zu einer Metaphorologie [Paradigms for a Metaphorology], Frankfurt am Main: Suhrkamp.

Blumenberg, Hans ([1960] 2010): Paradigms for a Metaphorology, Ithaca: Cornell University Press.

Blumer, Herbert (1969): Symbolic Interactionism. Perspective and Method, Englewood Cliffs: Prentice-Hall.

Bogner, Alexander/Littig, Beate/Menz, Wolfgang (Eds.) (2009): Interviewing Experts, Basingstoke, London: Palgrave Macmillan.

Bogner, Alexander/Menz, Wolfgang (2009): »The Theory-Generating Expert Interview. Epistemological Interest, Forms of Knowledge, Interaction«, in: Bogner/Littig/Menz (Eds.), Interviewing Experts, pp. 43–81.

Bogner, Alexander/Menz, Wolfgang/Littig, Beate (2009): »Introduction. Expert Interviews – An Introduction to a New Methodological Debate«, in: Interviewing Experts, pp. 1–13.

Böhning-Gaese, Katrin/Bauer, Friederike (2022): »The Loss of Biodiversity is Destroying the very Foundations of our Existence«. Interview. Akzente, https://akzente.giz.de/en/cop15-biodiversity-boehning-gaese from July 10, 2025.

Bonneuil, Christophe (2019): »Seeing Nature as a ›Universal Store of Genes‹. How Biological Diversity became ›Genetic Resources‹, 1890–1940«, in: Studies in History and Philosophy of Biological and Biomedical Sciences 75, pp. 1–14.

Bonneuil, Christophe/Fressoz, Jean-Baptiste (2017): The Shock of the Anthropocene. The Earth, History and Us, London, New York: Verso.

Boschen, Marleen (2022): Carrier Seeds. A Cultural Analysis of Care and Conflict in Four Seed Banking Practices. Thesis for the Degree of Doctor of Philosophy, Department of Media, Communications and Cultural Studies, Goldsmiths, London.

Bowker, Geoffrey C. (2000): »Biodiversity Datadiversity«, in: Social Studies of Science 30 (5), pp. 643–683.
Bowker, Geoffrey C. (2005): Memory Practices in the Sciences, Cambridge, MA: MIT Press.
Bowker, Geoffrey C./Star, Susan L. (1999): Sorting Things Out. Classification and its Consequences, Cambridge, MA: MIT Press.
Bowker, Geoffrey C./Star, Susan L. (2007): »Enacting Silence. Residual Categories as a Challenge for Ethics, Information Systems, and Communication«, in: Ethics and Information Technology 9 (4), pp. 273–280.
Braun, Sebastian F. (2017): »Mourning Ourselves and/as Our Relatives. Environment as Kinship«, in: Cunsolo/Landman (Eds.), Mourning Nature, pp. 64–91.
Braun, Veit/Lafuente-Funes, Sara/Lemke, Thomas/Liburkina, Ruzana (2023): »Making Futures by Freezing Life. Ambivalent Temporalities of Cryopreservation Practices«, in: Science, Technology, & Human Values 48 (4), pp. 693–699.
Breen, Sheryl (2015): »Saving Seeds. The Svalbard Global Seed Vault, Native American Seed-Savers, and Problems of Property«, in: Journal of Agriculture, Food Systems, and Community Development 5 (2), pp. 39–52.
Breithoff, Esther (2020): »Proxies«, in: Harrison/DeSilvey/Holtorf et al. (Eds.), Heritage Futures, pp. 121–131.
Breithoff, Esther/Harrison, Rodney (2020): »Banking Time. Trading in Futures«, in: Harrison/DeSilvey/Holtorf et al. (Eds.), Heritage Futures, pp. 101–120.
Bröckling, Ulrich (2023): »Apokalyptische Gouvernementalität. Eine Skizze« [»Apocalyptic Governmentality. A Draft«], in: Hoppe/Rüppel/von Verschuer/Voigt (Eds.), Leben Regieren [Governing Life], pp. 55–70.
Brockway, Lucile H. (1988): »Plant Science and Colonial Expansion. The Botanical Chess Game«, in: Kloppenburg (Ed.), Seeds and Sovereignty, pp. 49–66.
Brockway, Lucile H. (2002): Science and Colonial Expansion. The Role of the British Royal Botanic Gardens, New Haven: Yale University Press.
Brockway, Lucile H. (2011): »Science and Colonial Expansion. The Role of the British Royal Botanic Gardens«, in: Harding (Ed.), The Postcolonial Science and Technology Studies Reader, pp. 127–139.
Bulletin of the Atomic Scientists (2025): Doomsday Clock, https://thebulletin.org/doomsday-clock/ from July 16, 2025.
Burton-Christie, Douglas (2011): »The Gift of Tears. Loss, Mourning and the Work of Ecological Restoration«, in: Worldviews 15 (1), pp. 29–46.

Butler, Judith (2004): Precarious Life. The Powers of Mourning and Violence, London: Verso.

Butler, Judith (2020): The Force of Nonviolence. An Ethico-Political Bind, London, New York: Verso.

Catalyst Journal Editorial Board (2015): »Introduction to the Inaugural Issue«, in: Catalyst: Feminism, Theory, Technoscience 1 (1), pp. 1–11.

Catton, William R./Dunlap, Riley E. (1978): »Environmental Sociology. A New Paradigm«, in: The American Sociologist 13 (1), pp. 41–49.

Ceballos, Gerardo/Ehrlich, Paul R./Barnosky, Anthony D. et al. (2015): »Accelerated Modern Human-Induced Species Losses. Entering the Sixth Mass Extinction«, in: Science Advances 1 (5), pp. 1–5.

Chacko, Xan S. (2019a): »Creative Practices of Care. The Subjectivity, Agency, and Affective Labor of Preparing Seeds for Long-Term Banking«, in: Culture, Agriculture, Food and Environment 41 (2), pp. 97–106.

Chacko, Xan S. (2019b): »Digging Up Colonial Roots. The Less-Known Origins of the Millennium Seed Bank Partnership«, in: Catalyst: Feminism, Theory, Technoscience 5 (2), pp. 1–9.

Chacko, Xan S. (2022a): »Invisible Vitality. The Hidden Labours of Seed Banking«, in: Jenny Bangham/Xan S. Chacko/Judith Kaplan (Eds.), Invisible Labour in Modern Science, Lanham: Rowman & Littlefield, pp. 217–226.

Chacko, Xan S. (2022b): »Stringing, Reconnecting, and Breaking the Colonial ›Daisy Chain‹. From Botanic Garden to Seed Bank«, in: Catalyst: Feminism, Theory, Technoscience 8 (1), pp. 1–30.

Chao, Sophie (2023): »Multispecies Mourning. Grieving as Resistance on the West Papuan Plantation Frontier«, in: Cultural Studies 37 (4), pp. 553–579.

Chao, Sophie/Enari, Dion (2021): »Decolonising Climate Change. A Call for Beyond-Human Imaginaries and Knowledge Generation«, in: eTropic: Electronic Journal of Studies in the Tropics 20 (2), pp. 32–54.

Charmaz, Kathy (2006): Constructing Grounded Theory. A Practical Guide Through Qualitative Analysis, Los Angeles: SAGE.

Charmaz, Kathy (2014): Constructing Grounded Theory, Los Angeles, London, New Delhi, Singapore, Washington DC: SAGE.

Chwałczyk, Franciszek (2020): »Around the Anthropocene in Eighty Names. Considering the Urbanocene Proposition«, in: Sustainability 12, pp. 1–33.

Clarke, Adele E. (2005): Situational Analysis. Grounded Theory after the Postmodern Turn, Thousand Oaks: SAGE Publications.

Clarke, Adele E. (2015): »From Grounded Theory to Situational Analysis. What's New? Why? How?«, in: Clarke/Friese/Washburn (Eds.), Situational Analysis in Practice, pp. 84–118.

Clarke, Adele E./Friese, Carrie/Washburn, Rachel S. (2015a): »Part I: Introducing Situational Analysis«, in: Situational Analysis in Practice, pp. 11–75.

Clarke, Adele E./Friese, Carrie/Washburn, Rachel S. (2015b): »Part III: Exemplars of Situational Analysis Research. Introduction«, in: Situational Analysis in Practice, pp. 171–194.

Clarke, Adele E./Friese, Carrie/Washburn, Rachel S. (2018): Situational Analysis. Grounded Theory after the Interpretive Turn, Los Angeles: SAGE.

Clarke, Adele E./Friese, Carrie/Washburn, Rachel S. (Eds.) (2015c): Situational Analysis in Practice. Mapping Research with Grounded Theory, Walnut Creek: Left Coast Press.

Clarke, Adele E./Washburn, Rachel S./Friese, Carrie (Eds.) (2022): Situational Analysis in Practice. Mapping Relationalities Across Disciplines, New York: Routledge.

Clewell, Tammy (2004): »Mourning Beyond Melancholia. Freud's Psychoanalysis of Loss«, in: Journal of the American Psychoanalytic Association 52 (1), pp. 43–67.

Cooper, David/Engels, Jan/Frison, Emile (1994): A Multilateral System for Plant Genetic Resources. Imperatives, Achievements and Challenges, Rome: International Plant Genetic Resources Institute (IPGRI).

Corbin, Juliet M./Strauss, Anselm L. (2015): Basics of Qualitative Research. Techniques and Procedures for Developing Grounded Theory, Thousand Oaks: SAGE.

Cowie, Robert H./Bouchet, Philippe/Fontaine, Benoît (2022): »The Sixth Mass Extinction. Fact, Fiction or Speculation?«, in: Biological Reviews of the Cambridge Philosophical Society 97, pp. 640–663.

Crop Trust (2011): Sacred Valley of the Incas to Send Potatoes to Seed Vault, Cusco, Peru.

Crop Trust (2015): Andean Potato Varieties Once Lost, Now Safeguarded Forever, Svalbard, Norway.

Crop Trust (2019): The Plant Treaty. Q&A with Crop Trust Director of Science, Luigi Guarino, https://www.croptrust.org/spotlight/the-plant-treaty/ from June 15, 2025.

Crop Trust (2020): Seed Summit Highlights Need for Genetic Diversity, https://www.croptrust.org/news-events/news/seed-summit-highlights-need-for-genetic-diversity/ from October 29, 2025.

Crop Trust (2022): What is a Genebank?, https://www.croptrust.org/news-events/news/what-is-a-genebank/ from June 15, 2025.

Crop Trust (2025): Endowment Fund, https://www.croptrust.org/what-we-do/endowment-fund/ from July 18, 2025.

Crosby, Alfred W. (1972): The Columbian Exchange. Biological and Cultural Consequences of 1492, Westport, CT: Greenwood Press.

Crutzen, Paul J. (2002): »Geology of Mankind«, in: Nature 415 (6867), p. 23.

Crutzen, Paul J./Stoermer, Eugene F. (2000): »The ›Anthropocene‹«, in: Global Change Newsletter 41, pp. 17–18.

Cunsolo, Ashlee (2017): »Climate Change as the Work of Mourning«, in: Cunsolo/Landman (Eds.), Mourning Nature, pp. 169–189.

Cunsolo, Ashlee/Ellis, Neville R. (2018): »Ecological Grief as a Mental Health Response to Climate Change-Related Loss«, in: Nature Climate Change 8 (4), pp. 275–281.

Cunsolo, Ashlee/Landman, Karen (2017a): »Introduction. To Mourn Beyond the Human«, in: Mourning Nature, pp. 3–26.

Cunsolo, Ashlee/Landman, Karen (Eds.) (2017b): Mourning Nature. Hope at the Heart of Ecological Loss and Grief, Montreal, Kingston, London, Chicago: McGill-Queen's University Press.

Curry, Helen A. (2017a): »Breeding Uniformity and Banking Diversity. The Genescapes of Industrial Agriculture, 1935–1970«, in: Global Environment 10 (1), pp. 83–113.

Curry, Helen A. (2017b): »From Working Collections to the World Germplasm Project. Agricultural Modernization and Genetic Conservation at the Rockefeller Foundation«, in: History and Philosophy of the Life Sciences 39 (5), pp. 1–20.

Curry, Helen A. (2019a): »Gene Banks, Seed Libraries, and Vegetable Sanctuaries. The Cultivation and Conservation of Heritage Vegetables in Britain, 1970–1985«, in: Culture, Agriculture, Food and Environment 41 (2), pp. 87–96.

Curry, Helen A. (2019b): »Introduction: The Collection and Conservation of Plant Genetic Resources, Past and Present«, in: Culture, Agriculture, Food and Environment 41 (2), pp. 73–75.

Curry, Helen A. (2022a): Endangered Maize. Industrial Agriculture and the Crisis of Extinction, Berkeley: University of California Press.

Curry, Helen A. (2022b): »The History of Seed Banking and the Hazards of Backup«, in: Social Studies of Science 52 (5), pp. 664–688.

Dalyan, Can (2021): »Latent Capital. Seed Banking as Investment in Climate Change Futures«, in: Marion Ernwein/Franklin Ginn/James Palmer (Eds.), The Work That Plants Do. Life, Labour and the Future of Vegetal Economies, Bielefeld: transcript, pp. 181–192.

Danowski, Déborah/Viveiros de Castro, Eduardo (2017): The Ends of the World, Cambridge, Malden, MA: Polity Press.

Dányi, Endre (2020): »Búskomor Politics. Practising Critique in the Ruins of Liberal Democracy«, in: The Sociological Review 68 (2), pp. 356–368.

Dányi, Endre (2022): Melancholy Democracy. Politics Beyond Hope and Despair. Habilitation, Department of Social Sciences, Goethe University Frankfurt, Frankfurt am Main.

de la Cadena, Marisol/Blaser, Mario (2018b): »Pluriverse. Proposals for a World of Many Worlds«, in: A World of Many Worlds, pp. 1–22.

de la Cadena, Marisol/Blaser, Mario (Eds.) (2018a): A World of Many Worlds, Durham: Duke University Press.

Dempewolf, Hannes (2020): »Why the Seed Vault Matters to Genebanks«, The 2020 Svalbard Seed Summit. Genetic Diversity for more Resilient Food Systems, Longyearbyen, Svalbard, 25 February.

Descola, Philippe (1994): In the Society of Nature. A Native Ecology in Amazonia, Cambridge, UK: Cambridge University Press.

Descola, Philippe (2005): »Ecology as Cosmological Analysis«, in: Alexandre Surrallés/Pedro G. Hierro (Eds.), The Land Within. Indigenous Territory and Perception of Environment, Copenhagen: International Work Group for Indigenous Affairs, pp. 22–35.

Descola, Philippe (2014): Beyond Nature and Culture, Chicago, London: The University of Chicago Press.

Desmond, Matthew (2014): »Relational Ethnography«, in: Theory and Society 43 (5), pp. 547–579.

Dirzo, Rodolfo/Raven, Peter H. (2003): »Global State of Biodiversity and Loss«, in: Annual Review of Environment and Resources 28 (1), pp. 137–167.

Dodds, Joseph (2011): Psychoanalysis and Ecology at the Edge of Chaos. Complexity Theory, Deleuze/Guattari and Psychoanalysis for a Climate in Crisis, London, New York: Routledge.

Doka, Kenneth J. (Ed.) (1989): Disenfranchised Grief. Recognizing Hidden Sorrow, Lexington, MA: Lexington Press.

Dunlap, Riley E./Catton, William R. (1979): »Environmental Sociology«, in: Annual Review of Sociology 5, pp. 243–273.

Durkheim, Émile ([1912] 2008): The Elementary Forms of Religious Life, Oxford: Oxford University Press.

Engert, Kornelia/Schürkmann, Christiane (2021): »Introduction. Posthuman? Nature and Culture in Renegotiation«, in: Nature and Culture 16 (1), pp. 1–10.

Escobar, Arturo (1999): »After Nature. Steps to an Antiessentialist Political Ecology«, in: Current Anthropology 40 (1), pp. 1–30.

Escobar, Arturo (2018): Designs for the Pluriverse. Radical Interdependence, Autonomy, and the Making of Worlds, Durham, London: Duke University Press.

Escobar, Arturo (2020): Pluriversal Politics. The Real and the Possible, Durham, London: Duke University Press.

Esquinas-Alcázar, José/Hilmi, Angela/López Noriega, Isabel (2012): »A Brief History of the Negotiations on the International Treaty on Plant Genetic Resources for Food and Agriculture«, in: Halewood/López Noriega/Louafi (Eds.), Crop Genetic Resources as a Global Commons, pp. 135–149.

ETC Group (2001): »The Law of the Seed!«, in: ETC Translator 3 (1), pp. 1–16.

ETC Group (n.d.): ETC Group. A Brief History, https://www.etcgroup.org/content/etc-group-brief-history from July 10, 2025.

EU Observer (2008): »Barroso to open ›Noah's Ark‹ for Seeds«, from February 22, written by Lisbeth Kirk, https://euobserver.com/news/25710 from July 16, 2025.

Faier, Lieba/Rofel, Lisa (2014): »Ethnographies of Encounter«, in: Annual Review of Anthropology 43, pp. 363–377.

FAO (1983): International Undertaking on Plant Genetic Resources, https://lib.icimod.org/record/9792/files/344.pdf from June 18, 2025.

FAO (1989): Resolution 4/89. Agreed Interpretation of the International Undertaking, https://www.fao.org/4/x5588E/x5588e06.htm#Resolution4 from June 18, 2025.

FAO (1991): Resolution 3/91. Annex 3 to the International Undertaking on Plant Genetic Resources, https://www.fao.org/4/x5587E/x5587e06.htm#Resolution3 from June 18, 2025.

FAO (1996a): Global Plan of Action for the Conservation and Sustainable Utilization of Plant Genetic Resources for Food and Agriculture and the Leipzig Declaration. Adopted by the International Technical Conference on Plant Genetic Resources, Leipzig, Germany, 17–23 June 1996, Rome, https://openknowledge.fao.org/server/api/core/bitstreams/8acc690b-246f-40b9-948d-c85070bc6a0c/content from July 25, 2025.

FAO (1996b): Report of the International Technical Conference on Plant Genetic Resources. Leipzig, Germany, 17–23 June 1996, Document: ITCPGR/96/Rep, Rome.

FAO (1997): The State of the World's Plant Genetic Resources for Food and Agriculture, Rome, https://openknowledge.fao.org/server/api/core/bitstreams/0172f037-db3b-47e8-9f93-26a1701db513/content from July 25, 2025.

FAO (2001): International Treaty on Plant Genetic Resources for Food and Agriculture. Opened for Signature on November 03, 2001, Rome, https://openknowledge.fao.org/server/api/core/bitstreams/a9d0de2a-8e98-4f75-98a8-673078841030/content from July 25, 2025.

FAO (2006): Relationship Agreement Between the Global Crop Diversity Trust and the Governing Body of the International Treaty on Plant Genetic Resources for Food and Agriculture, Madrid, https://www.croptrust.org/fileadmin/uploads/croptrust/Documents/Policy_Documents/Signed-Relationship-Agreement_0.pdf from July 16, 2025.

FAO (2010): The Second Report on the State of the World's Plant Genetic Resources for Food and Agriculture, Rome, https://www.fao.org/4/i1500e/i1500e00.htm from July 25, 2025.

FAO (2011): Second Global Plan of Action for Plant Genetic Resources for Food and Agriculture. Adopted by the FAO Council, Rome, Italy, 29 November 2011, Rome, https://www.fao.org/4/i2624e/i2624e00.htm from July 25, 2025.

FAO (2014): Genebank Standards for Plant Genetic Resources for Food and Agriculture, Rome https://www.fao.org/4/i3704e/i3704e.pdf from July 25, 2025.

FAO (2019): The State of the World's Biodiversity for Food and Agriculture, Rome, https://openknowledge.fao.org/server/api/core/bitstreams/50b79369-9249-4486-ac07-9098d07df60a/content from July 25, 2025.

FAO (2025): List of Contracting Parties. International Treaty on Plant Genetic Resources for Food and Agriculture, https://www.fao.org/plant-treaty/countries/contracting-parties/en from September 29, 2025.

Fava, Sergio (2013): Environmental Apocalypse in Science and Art. Designing Nightmares, New York: Routledge.

Fenzi, Marianna/Bonneuil, Christophe (2016): »From ›Genetic Resources‹ to ›Ecosystems Services‹. A Century of Science and Global Policies for Crop Diversity Conservation«, in: Culture, Agriculture, Food and Environment 38 (2), pp. 72–83.

Flick, Uwe (2014): An Introduction to Qualitative Research, Los Angeles: SAGE.

Flitner, Michael (1995): Sammler, Räuber und Gelehrte. Die politischen Interessen an pflanzengenetischen Ressourcen 1895–1995 [Collectors, Raiders and Scholars. Political Interests in Plant Genetic Resources 1895–1995], Frankfurt am Main: Campus.

Flitner, Michael (2003): »Genetic Geographies. A Historical Comparison of Agrarian Modernization and Eugenic Thought in Germany, the Soviet Union, and the United States«, in: Geoforum 34 (2), pp. 175–185.

Flitner, Michael/Görg, Christoph/Heins, Volker (Eds.) (1998): Konfliktfeld Natur. Biologische Ressourcen und globale Politik [Nature as an Area of Conflict. Biological Resources and Global Politics], Opladen: Leske + Budrich.

Flitner, Michael/Heins, Volker (2002): »Modernity and Life Politics. Conceptualizing the Biodiversity Crisis«, in: Political Geography 21 (3), pp. 319–340.

Flores-Palacios, Ximena (1998): Contribution to the Estimation of Countries' Interdependence in the Area of Plant Genetic Resources. Background Study Paper No. 7, Rev. 1, Rome.

Folkers, Andreas (2017): »Politik des Lebens jenseits seiner selbst. Für eine ökologische Lebenssoziologie mit Deleuze und Guattari« [»The Politics of Life Beyond Itself. For an Ecological Sociology of Life with Deleuze and Guattari«], in: Soziale Welt 68 (4), pp. 365–384.

Foucault, Michel (1970): The Order of Things. An Archaeology of the Human Sciences, London: Tavistock.

Foucault, Michel (1978): The History of Sexuality. Vol. 1: An Introduction, New York: Pantheon Books.

Foucault, Michel (Ed.) (2003): Society must be Defended. Lectures at the Collège de France, 1975–76, New York: Picador.

Fowler, Cary (1994): Unnatural Selection. Technology, Politics, and Plant Evolution, Yverdon: Gordon and Breach.

Fowler, Cary (2008a): »Crop Diversity. Neolithic Foundations for Agriculture's Future Adaptation to Climate Change«, in: Ambio Special Report Number 14, pp. 498–501.

Fowler, Cary (2008b): »The Svalbard Seed Vault and Crop Security«, in: BioScience 58 (3), pp. 190–191.

Fowler, Cary (2008c): The Svalbard Global Seed Vault. Securing the Future of Agriculture, https://farm-d.org/document/the-svalbard-global-seed-vault-securing-the-future-of-global-agriculture/ from July 16, 2025.

Fowler, Cary (2009): »One Seed at a Time. Protecting the Future of Food«, TED Global, https://www.ted.com/talks/cary_fowler_one_seed_at_a_time_protecting_the_future_of_food/transcript?language=en from June 10, 2024.

Fowler, Cary (2016): Seeds on Ice. Svalbard and the Global Seed Vault, Westport, CT: Prospecta Press.

Fowler, Cary/George, William/Shands, Henry et al. (2004): »Study to Assess the Feasibility of Establishing a Svalbard Arctic Seed Depository for the International Community. Prepared for the Ministry of Foreign Affairs and the Ministry of Agriculture.« Center for International Environment and Development Studies (Noragric), Agricultural University of Norway & Nordic Gene Bank. https://www.regjeringen.no/globalassets/upload/lmd/kampanjesvalbard/vedlegg/frohvelv_study_to_assess.pdf from June 12, 2024.

Fowler, Cary/Hodgkin, Toby (2004): »Plant Genetic Resources for Food and Agriculture. Assessing Global Availability«, in: Annual Review of Environment and Resources 29 (1), pp. 143–179.

Fowler, Cary/Mooney, Pat R. (1990): Shattering. Food, Politics, and the Loss of Genetic Diversity, Tucson: The University of Arizona Press.

Frankel, Otto H./Agble, W. K./Harlan, Jack R./Bennett, Erna ([1969] 1970): »Genetic Dangers in the Green Revolution«, in: The Ecologist 1 (4), pp. 20–22.

Frankel, Otto H./Bennett, E. (Eds.) (1970): Genetic Resources in Plants. Their Exploration and Conservation, Oxford: Blackwell Scientific.

Franklin, Sarah (2000): »Life Itself. Global Nature and the Genetic Imaginary«, in: Sarah Franklin/Celia Lury/Jackie Stacey (Eds.), Global Nature, Global Culture, London: SAGE Publications, pp. 188–227.

Freud, Sigmund ([1917] 1957): »Mourning and Melancholia«, in: The Standard Edition of the Complete Psychological Works of Sigmund Freud. Translated from German under the General Editorship of James Strachey, in Collaboration with Anna Freud, London: The Hogarth Press, pp. 238–258.

Friedrich, Alexander (2017): »The Rise of Cryopower. Biopolitics in the Age of Cryogenic Life«, in: Kowal/Radin (Eds.), Cryopolitics, 59–69.

Friedrich, Alexander (2020): »A Cold Yield. Cryopreserved Oocytes of ›Social Freezing‹ Customers as Potential Option Values for Biomedical Research«, in: New Genetics and Society 39 (3), pp. 327–351.

Friedrich, Alexander/Höhne, Stefan (2016): »Regimes of Freshness. Biopolitics in the Age of Cryogenic Culture«, in: Medicine Anthropology Theory 3 (3), pp. 112–154.

Frost, Samantha (2016): Biocultural Creatures. Toward a New Theory of the Human, Durham: Duke University Press.

Fu, Yong-Bi (2017): »The Vulnerability of Plant Genetic Resources Conserved Ex Situ«, in: Crop Science 57 (5), pp. 2314–2328.

Fullilove, Courtney (2017): The Profit of the Earth. The Global Seeds of American Agriculture, Chicago: University of Chicago Press.

Gaard, Greta (2015): »Ecofeminism and Climate Change«, in: Women's Studies International Forum 49, pp. 20–33.

Gan, Elaine (2015): »Seed Vault. Freezing Life for Doomsday«, in: James Brady (Ed.), Elemental. An Arts and Ecology Reader, Manchester, UK: Gaia Project Press, pp. 119–121.

Gan, Elaine (2017): »An Unintended Race. Miracle Rice and the Green Revolution«, in: Environmental Philosophy 14 (1), pp. 61–81.

Gan, Elaine (2019): »Sorting Seeds into Racialized Futures and Pasts«, in: Catalyst: Feminism, Theory, Technoscience 5 (2), pp. 1–3.

Garcés, Marina/Hentschel, Christine (2023): »›As If We Have Not Already Destroyed Many Worlds Before‹. Marina Garcés in Conversation with Christine Hentschel«, in: Soziopolis: Gesellschaft beobachten, https://www.ssoar.info/ssoar/handle/document/90607 from June 12, 2024.

GenØk – Centre for Biosafety (Ed.) (2019): Agri/Cultures.Seed-Links Exhibition. Poster, https://genok.com/arkiv/9614/ from March 20, 2024.

Gibbon, Sahra/Novas, Carlos (Eds.) (2008): Genetics, Biosociality, and the Social Sciences. Making Biologies and Identities, Abingdon, New York: Routledge.

Goffman, Erving ([1967] 1989): Interaction Ritual. Essays on Face-to-Face Behavior, New York: Pantheon Books.

Gómez-Baggethun, Erik (2019): »Sustainable Development«, in: Kothari/Salleh/Escobar et al. (Eds.), Pluriverse, pp. 71–74.

Gómez-Barris, Macarena (2017): The Extractive Zone. Social Ecologies and Decolonial Perspectives, Durham, London: Duke University Press.

GRAIN (2001): »A Disappointing Compromise«, in: Seedling 18 (4), https://grain.org/en/article/329-a-dissapointing-compromise from June 10, 2024.

GRAIN (2008): Faults in the Vault. Not Everyone is Celebrating Svalbard, https://www.grain.org/fr/article/entries/181-faults-in-the-vault-not-everyone-is-celebrating-svalbard from June 18, 2025.

GRAIN (n.d.): Organisation, https://grain.org/en/pages/organisation from July 10, 2025.

Grigoli, Rebecca (2024): »Planting the Seed. The Svalbard Global Seed Vault as an Archive in the Face of Uncertain Futures«, in: The iJournal 9 (2), pp. 117–131.

Grillmayr, Julia/Hentschel, Christine (2024): »World Without Humans, Humans Without World. Apocalyptic Passions in the Anthropocene«, in: Jenny Stümer/Michael Dunn (Eds.), Worlds Ending. Ending Worlds. Understanding Apocalyptic Transformation, Basel, Berlin, Boston: De Gruyter, pp. 209–226.

Grimes, Ronald L. (1982): Beginnings in Ritual Studies, Washington, D.C: University Press of America.

Guarino, Luigi (2018): The Art of the Multilateral Deal. Science Blog, Crop Trust, https://www.croptrust.org/science-blog/art-of-the-multilateral-deal/ from June 18, 2025.

Gullion, Jessica S. (2016): Writing Ethnography, Rotterdam: SensePublishers.

Gutiérrez Rodriguez, Encarnación (2023): Decolonial Mourning and the Caring Commons. Migration-Coloniality Necropolitics and Conviviality Infrastructure, London, New York: Anthem Press.

Halewood, Michael/López Noriega, Isabel/Louafi, Sélim (Eds.) (2012): Crop Genetic Resources as a Global Commons. Challenges in International Law and Governance, London: Taylor and Francis.

Handwerk, Brian (2012): »Pictures: ›Doomsday‹ Seed Vault Safeguards Our Food Supply«, in: National Geographic from July 5, 2012, https://www.nationalgeographic.com/culture/article/120702-svalbard-doomsday-seed-vault-food-supply from July 10, 2025.

Hanelt, Peter/Knüpffer, Helmut/Hammer, Karl (2012): »Erna Bennett (5 August 1925–3 January 2012)«, in: Genetic Resources and Crop Evolution 59 (6), pp. 967–970.

Haraway, Donna J. (1988): »Situated Knowledges: The Science Question in Feminism and the Privilege of Partial Perspective«, in: Feminist Studies 14 (3), pp. 575–599.

Haraway, Donna J. (1989): Primate Visions. Gender, Race, and Nature in the World of Modern Science, New York: Routledge.

Haraway, Donna J. (1991): Simians, Cyborgs, and Women. The Reinvention of Nature, New York: Routledge.

Haraway, Donna J. (1992): »The Promises of Monsters. A Regenerative Politics for Inappropriate/d Others«, in: Lawrence Grossberg (Ed.), Cultural Studies, New York: Routledge, pp. 295–337.

Haraway, Donna J. (1997): Modest_Witness@Second_Millennium. FemaleMan©_Meets_OncoMouse™. Feminism and Technoscience, London: Routledge.

Haraway, Donna J. (2003): The Companion Species Manifesto. Dogs, People, and Significant Otherness, Chicago: Prickly Paradigm Press.

Haraway, Donna J. (2008): When Species Meet, Minneapolis: University of Minnesota Press.

Haraway, Donna J. (2016): Staying with the Trouble. Making Kin in the Chthulucene, Durham: Duke University Press.

Haraway, Donna J. (2023): »Present to Bruno, from Donna«, in: Social Studies of Science 53 (2), pp. 165–168.

Harding, Sandra G. (1986): The Science Question in Feminism, Ithaca: Cornell University Press.

Harding, Sandra G. (1991): Whose Science? Whose Knowledge? Thinking from Women's Lives, Ithaca: Cornell University Press.

Harding, Sandra G. (Ed.) (2011): The Postcolonial Science and Technology Studies Reader, Durham, NC: Duke University Press.

Harlan, Jack R. (1972): »Genetics of Disaster«, in: Journal of Environmental Quality 1 (3), pp. 212–215.

Harlan, Jack R. (1975): »Our Vanishing Genetic Resources. Modern Varieties Replace Ancient Populations that have Provided Genetic Variability for Plant Breeding Programs«, in: Science 188, pp. 618–621.

Harlan, Jack R. (1988): »Seeds and Sovereignty. An Epilogue«, in: Kloppenburg (Ed.), Seeds and Sovereignty, pp. 356–362.

Harrison, Rodney (2015): »Beyond ›Natural‹ and ›Cultural‹ Heritage. Toward an Ontological Politics of Heritage in the Age of Anthropocene«, in: Heritage & Society 8 (1), pp. 24–42.

Harrison, Rodney (2017): »Freezing Seeds and Making Futures. Endangerment, Hope, Security, and Time in Agrobiodiversity Conservation Practices«, in: Culture, Agriculture, Food and Environment 39 (2), pp. 80–89.

Harrison, Rodney/DeSilvey, Caitlin/Holtorf, Cornelius et al. (Eds.) (2020): Heritage Futures. Comparative Approaches to Natural and Cultural Heritage Practices, London: UCL Press.

Hartigan, John (2017): Care of the Species. Races of Corn and the Science of Plant Biodiversity, Minneapolis: University of Minnesota Press.

Hawthorne, Susan (2007): »Land, Bodies, and Knowledge. Biocolonialism of Plants, Indigenous Peoples, Women, and People with Disabilities«, in: Signs: Journal of Women in Culture and Society 32 (2), pp. 314–323.

Head, Lesley (2016a): »Grief, Loss and the Cultural Politics of Climate Change«, in: Harriet Bulkeley/Johannes Stripple/Matthew Paterson (Eds.), Towards a Cultural Politics of Climate Change. Devices, Desires, and Dissent, Cambridge, UK: Cambridge University Press, pp. 81–93.

Head, Lesley (2016b): Hope and Grief in the Anthropocene. Re-Conceptualising Human-Nature Relations, London, New York: Routledge.

Heatherington, Tracey/Perley, Bernard C. (2017): »Fieldnotes from Svalbard. How Global Dreamings Take Root in the Arctic Frontier«, in: EuropeNow: A Journal of Research & Art. https://www.europenowjournal.org/2017/05/02/fieldnotes-from-svalbard-how-global-dreamings-take-root-in-the-arctic-frontier/ from June 10, 2024.

Heins, Volker/Flitner, Michael (1998): »Biologische Ressourcen und ›Life Politics‹« [Biological Resources and ›Life Politics‹], in: Flitner/Görg/Heins (Eds.), Konfliktfeld Natur [Nature as an Area of Conflict], pp. 13–38.

Helmreich, Stefan (2011): »What Was Life? Answers from Three Limit Biologies«, in: Critical Inquiry 37 (4), pp. 671–696.

Higgins, David (2021): »Apocalypse/Extinction«, in: Jeffrey J. Cohen/Stephanie Foote (Eds.), The Cambridge Companion to Environmental Humanities, Cambridge, UK: Cambridge University Press, pp. 114–127.

Hill Collins, Patricia ([1990] 2000): Black Feminist Thought. Knowledge, Consciousness, and the Politics of Empowerment, New York, London: Routledge.

Hoeyer, Klaus (2017): »Suspense. Reflections on Cryopolitics of the Body«, in: Kowal/Radin (Eds.), Cryopolitics, pp. 205–214.

Hopkins, Terence/Wallerstein, Immanuel M./Kasaba, Reşat et al. (1987): »Incorporation into the World-Economy. How the World-System Expands«, in: Review (Fernand Braudel Center) 10 (5/6), pp. 761–902.

Hoppe, Katharina (2024): »Dependency Denial in Crisis. Revisiting Feminist Critiques of Dualism«, in: European Journal of Social Theory 28 (1), pp. 68–84.

Hoppe, Katharina/Rüppel, Jonas/von Verschuer, Franziska/Voigt, Torsten H. (Eds.) (2023): Leben Regieren. Natur, Technologie und Gesellschaft im 21. Jahrhundert [Governing Life. Nature, Technology, and Society in the 21st Century], Frankfurt am Main: Campus.

Hubert, Henri/Mauss, Marcel ([1898] 1981): Sacrifice. Its Nature and Function, Chicago: University of Chicago Press.

Humlum, Ole (2005): »Svalbard«, in: Nuttall (Ed.), Encyclopedia of the Arctic, pp. 1975–1982.

IBPGR (1991): Annual Report 1990, Rome.

Jackson, Zakiyyah I. (2020): Becoming Human. Matter and Meaning in an Antiblack World, New York: New York University Press.

Jameson, Fredric (1981): The Political Unconscious. Narrative as a Socially Symbolic Act, Ithaca: Cornell University Press.

Jarvis, Devra I./Hodgkin, Toby (Eds.) (1998): Strengthening the Scientific Basis of in situ Conservation of Agricultural Biodiversity on-farm. Options for Data Collecting and Analysis, Rome: IPGRI.

Jensen, An-Magritt (2009): »From Thailand to Svalbard. Migration on the Margins«, in: Asia Insights 1, pp. 13–15.

Kastler, Guy (2016): UPOV and the Seed Treaty. UPOV must Respect Farmer's Rights. Intervention at the UPOV Symposium, Brussels, https://viacampesina.org/en/upov-must-respect-farmers-rights/ from June 18, 2025.

Keller, Catherine (1996): Apocalypse Now and Then. A Feminist Guide to the End of the World, Boston: Beacon.

Keller, Catherine (2021): Facing Apocalypse. Climate, Democracy, and other Last Chances, Maryknoll: Orbis Books.

Khoury, Colin K./Achicanoy, Harold A./Bjorkman, Anne D. et al. (2015): Where Our Food Crops Come From. A New Estimation of Countries' Interdependence in Plant Genetic Resources. Policy Brief No. 25.

Khoury, Colin K./Achicanoy, Harold A./Bjorkman, Anne D. et al. (2016): »Origins of Food Crops Connect Countries Worldwide«, in: Proceedings of the Royal Society B: Biological Sciences 283 (1832), pp. 1–9.

Khoury, Colin K./Argumedo, Alejandro/Dempewolf, Hannes et al. (2019): With Diversity at the Heart. A 2050 Vision for Global Food System Transformation, https://hdl.handle.net/10568/100866 from July 10, 2025.

Khoury, Colin K./Brush, Stephen/Costich, Denise E. et al. (2022): »Crop Genetic Erosion. Understanding and Responding to Loss of Crop Diversity«, in: The New Phytologist 233 (1), pp. 84–118.

King, Barbara J. (2014): How Animals Grieve, Chicago: The University of Chicago Press.

King, Tiffany L. (2019): The Black Shoals. Offshore Formations of Black and Native Studies, Durham: Duke University Press.

Kirksey, Eben/Helmreich, Stefan (2010): »The Emergence of Multispecies Ethnography«, in: Cultural Anthropology 25 (4), pp. 545–576.

Kirksey, Eben/Schuetze, Craig/Helmreich, Stefan (2014): »Introduction. Tactics of Multispecies Ethnography«, in: Eben Kirksey (Ed.), The Multispecies Salon, Durham: Duke University Press, pp. 1–24.

Klein, Hilary (2015): Compañeras. Zapatista Women's Stories, New York: Seven Stories Press.

Kloppenburg, Jack R. ([1988] 2004): First the Seed. The Political Economy of Plant Biotechnology, 1492–2000, Madison: University of Wisconsin Press.

Kloppenburg, Jack R. (2000): »Biopiracy, Witchery, and the Fables of Ecoliberalism«, in: Peace Review 12 (4), pp. 509–516.

Kloppenburg, Jack R. (2014): »Re-Purposing the Master's Tools. The Open Source Seed Initiative and the Struggle for Seed Sovereignty«, in: The Journal of Peasant Studies 41 (6), pp. 1225–1246.

Kloppenburg, Jack R. (Ed.) (1988): Seeds and Sovereignty. The Use and Control of Plant Genetic Resources, Durham: Duke University Press.

Kloppenburg, Jack R./Kleinman, Daniel L. (1988): »Seeds of Controversy. National Property Versus Common Heritage«, in: Kloppenburg (Ed.), Seeds and Sovereignty, pp. 173–203.

Knowles, Scott G. (2014): »Learning from Disaster? The History of Technology and the Future of Disaster Research«, in: Technology and Culture 55 (4), pp. 773–784.

Kothari, Ashish/Salleh, Ariel/Escobar, Arturo et al. (Eds.) (2019): Pluriverse. A Post-Development Dictionary, New Delhi: Tulika Books.

Kowal, Emma/Radin, Joanna (Eds.) (2017): Cryopolitics. Frozen Life in a Melting World, Cambridge, MA: The MIT Press.

Kroløkke, Charlotte (2019): »Frosties. Feminist Cultural Analysis of Frozen Cells and Seeds Documentaries«, in: European Journal of Cultural Studies 22 (5–6), pp. 528–544.

Kropp, Cordula/Sonnberger, Marco (2025): Environmental Sociology, Baden-Baden: Nomos.

Kruse, Frigga (2012): »Spitsbergen – Imperialists Beyond the British Empire«, in: Louwrens Hacquebord (Ed.), LASHIPA. History of large-scale Resource Exploitation in Polar Areas, Eelde: Barkhuis Publishing, pp. 61–70.

Laboissière, Anna-Katharina (2019): »Collect, Save, Adapt. Making and Unmaking ex situ Worlds«, in: Cultural Studies Review 25 (1), pp. 65–84.

Latour, Bruno (2004a): »Whose Cosmos, Which Cosmopolitics? Comments on the Peace Terms of Ulrich Beck«. Symposium: Talking Peace with Gods, Part 1, in: Common Knowledge 10 (3), pp. 450–462.

Latour, Bruno (2004b): »Why Has Critique Run out of Steam? From Matters of Fact to Matters of Concern«, in: Critical Inquiry 30 (2), pp. 225–248.

Latour, Bruno (2013): An Inquiry into Modes of Existence. An Anthropology of the Moderns, Cambridge, MA: Harvard University Press.

Latour, Bruno (2017): Facing Gaia. Eight Lectures on the new Climate Regime, Cambridge, UK, Medford, MA: Polity Press.

Latour, Bruno (2018): Down to Earth. Politics in the New Climatic Regime, Cambridge, UK: Polity Press.

Latour, Bruno/Gertenbach, Lars/Opitz, Sven/Tellmann, Ute (2016): »›There is no Earth Corresponding to the Globe‹. An Interview with Bruno Latour«, in: Soziale Welt 67 (3), pp. 353–364.

Laurent, Anna (2019): Iraq. Recovering a Garden of Paradise. Seed Cultures Archive, https://www.seedcultures.com/#/anna-laurent/ from June 18, 2025.

Law, John (2015): »What's Wrong with a One-World World?«, in: Distinktion: Journal of Social Theory 16 (1), pp. 126–139.

Le Monde (2010): »Le Coffre-Fort de l'Apocalypse. L'Arche de Noë où Sauver Toutes les Semences du Monde« [»The Safe of the Apocalypse. Noah's Ark Where all the Seeds of the World are Saved«], from February 8, written by Frédéric Joignot, https://www.lemonde.fr/blog/fredericjoignot/2010/02/08/le-coffre-fort-de-lapocalypse/ from June 18, 2025.

Leinius, Johanna (2021): »Articulating Body, Territory, and the Defence of Life. The Politics of Strategic Equivalencing between Women in Anti-Mining Movements and the Feminist Movement in Peru«, in: Bulletin of Latin American Research 40 (2), pp. 204–219.

Lemke, Thomas (2012): »Second Nature. In the Age of Biobanks«, in: The Yearbook of Comparative Literature 58, pp. 188–192.

Lemke, Thomas (2021): »Welcome to Whenever. Exploring Suspended Life in Cryopreservation Practices«, in: Science, Technology, & Human Values 48 (4), pp. 700–726.

Lemke, Thomas (2022): »Conceptualising Suspended Life: From Latency to Liminality«, in: Theory, Culture & Society 40 (6), pp. 69–86.

Lewis, Simon L./Maslin, Mark A. (2015): »Defining the Anthropocene«, in: Nature Perspectives 519 (12 March), pp. 171–180.

Liboiron, Max (2021): Pollution is Colonialism, Durham, London: Duke University Press.

Lussier, Jessica (2021): »Relearning Our World. Grief, Loss, and Mourning-With«, in: Philosophy of Education 77, pp. 143–159.

Lykke, Nina (2019): »Co-Becoming with Diatoms«, in: Catalyst: Feminism, Theory, Technoscience 5 (2): pp. 1–25.

Lykke, Nina (2022): Vibrant Death. A Posthuman Phenomenology of Mourning, London: Bloomsbury.

Manzella, Daniele (2012): »The Design and Mechanics of the Multilateral System of Access and Benefit Sharing«, in: Halewood/López Noriega/Louafi (Eds.), Crop Genetic Resources as a Global Commons, pp. 150–163.

Marcus, George E. (1995): »Ethnography in/of the World System. The Emergence of Multi-Sited Ethnography«, in: Annual Review of Anthropology 24, pp. 95–117.

Marcus, George E. (1998): Ethnography Through Thick and Thin, Princeton: Princeton University Press.

Mark, Andrew/Di Battista, Amanda (2017): »Making Loss the Centre. Podcasting our Environmental Grief«, in: Cunsolo/Landman (Eds.), Mourning Nature, pp. 227–257.

Martin, Gary/Mincyte, Diana/Münster, Ursula (Eds.) (2012): Why Do We Value Diversity? Biocultural Diversity in a Global Context, Rachel Carson Center Perspectives 2012/9.

Masco, Joseph (2017): »The Crisis in Crisis«, in: Current Anthropology 58 (Supplement 15), pp. S65-S76.

Mbembe, Achille (2019): Necropolitics, Durham, London: Duke University Press.

McMichael, Philip (2019): »The Development Project«, in: Kothari/Salleh/Escobar et al. (Eds.), Pluriverse, pp. 12–15.

McMurry, Andrew (1996): »The Slow Apocalypse. A Gradualistic Theory of the World's Demise«, in: Postmodern Culture 6 (3), published online: https://doi.org/10.1353/pmc.1996.0018 from June 12, 2024.

Meek, Laura A./Morales Fontanilla, Julia A. (2022): »Otherwise«, in: Feminist Anthropology 3 (2), pp. 274–283.

Menning, Nancy (2017): »Environmental Mourning and the Religious Imagination«, in: Cunsolo/Landman (Eds.), Mourning Nature, pp. 39–63.

Merchant, Carolyn (1980): The Death of Nature. Women, Ecology, and the Scientific Revolution, San Francisco: Harper & Row.

Mignolo, Walter D. (2011): The Darker Side of Western Modernity. Global Futures, Decolonial Options, Durham: Duke University Press.

Mignolo, Walter D. (2015): »Sylvia Wynter. What does it Mean to be Human?«, in: Katherine McKittrick/Sylvia Wynter (Eds.), Sylvia Wynter. On Being Human as Praxis, Durham, London: Duke University Press, pp. 106–123.

Mignolo, Walter D. (2018): »The Decolonial Option. Part II«, in: Mignolo/Walsh (Eds.), On Decoloniality, pp. 103–244.

Mignolo, Walter D./Segato, Rita/Walsh, Catherine E. (Eds.) (2024): Aníbal Quijano. Foundational Essays on the Coloniality of Power, Durham: Duke University Press.
Mignolo, Walter D./Walsh, Catherine E. (Eds.) (2018): On Decoloniality. Concepts, Analytics, Praxis, Durham: Duke University Press.
Mihai, Mihaela/Thaler, Mathias (2023): »Environmental Commemoration. Guiding Principles and Real-World Cases«, in: Memory Studies 17 (6), pp. 1378–1395.
Mitchell, Audra/Chaudhury, Aadita (2020): »Worlding Beyond ›the‹ ›End‹ of ›the World‹. White Apocalyptic Visions and BIPOC Futurisms«, in: International Relations 34 (3), pp. 309–332.
Mitman, Gregg/Haraway, Donna J./Tsing, Anna L. (2019): »Reflections on the Plantationocene. A Conversation with Donna Haraway and Anna Tsing«, in: Edge Effects, https://edgeeffects.net/haraway-tsing-plantationocene/ from June 18, 2025.
Molina Vargas, Hema'ny/Marambio, Camila/Lykke, Nina (2020): »Decolonising Mourning. World-Making with the Selk'nam People of Karokynka/Tierra del Fuego«, in: Australian Feminist Studies 35 (104), pp. 186–201.
Moltmann, Jürgen (1967): Theology of Hope, New York: Harper & Row.
Moltmann, Jürgen (1996): The Coming of God. Christian Eschatology, Minneapolis: Fortress Press.
Mooney, Pat R. (1979): Seeds of the Earth. A Private or Public Resource? Ottawa: Inter Pares.
Mooney, Pat R. (1983): »The Law of the Seed – Another Development and Plant Genetic Resources«, in: Development Dialogue (1/2), pp. 1–172.
Moore, Jason W. (2000): »Sugar and the Expansion of the Early Modern World-Economy. Commodity Frontiers, Ecological Transformation, and Industrialization«, in: Review (Fernand Braudel Center) 23 (3), pp. 409–433.
Moore, Jason W. (2015): Capitalism in the Web of Life. Ecology and the Accumulation of Capital, London, New York: Verso.
Moore, Jason W. (Ed.) (2016): Anthropocene or Capitalocene? Nature, History, and the Crisis of Capitalism, Oakland: PM Press.
More Worlds Collective (Masco, Joseph/Choy, Timothy/Kosek, Jake/Murphy, M.) (2025): Fear of a Dead White Planet, Durham, London: Duke University Press.
Muñoz, José E. (1999): Disidentifications. Queers of Color and the Performance of Politics, Minneapolis: University of Minnesota Press.

Murphy, M. (2012): Seizing the Means of Reproduction. Entanglements of Feminism, Health, and Technoscience, Durham: Duke University Press.
Murphy, M. (2017): »Alterlife and Decolonial Chemical Relations«, in: Cultural Anthropology 32 (4), pp. 494–503.
Nadim, Tahani (2018): »Haunting Seedy Connections«, in: Celia Lury/Rachel Fensham/Alexandra Heller-Nicholas et al. (Eds.), Routledge Handbook of Interdisciplinary Research Methods, Georgetown: Routledge, pp. 239–247.
NASA (n.d.): »Arctic Sea Ice Minimum Extent«, in: Global Climate Change: Vital Signs of the Planet, https://climate.nasa.gov/vital-signs/arctic-sea-ice / from July 2, 2025.
Nath, Dipika (2009): »›To Abandon the Colonial Animal‹. ›Race‹, Animals, and the Feral Child in Kipling's Mowgli Stories«, in: Sarah E. McFarland/Ryan Hediger (Eds.), Animals and Agency. An Interdisciplinary Exploration, Leiden, Boston: Brill, pp. 251–277.
Nature (2008): »Snapshot. ›Doomsday Vault‹ Opens«, from February 28, written by Michael Hopkin, 452 (1037), n.p.
Nehring, Ryan (2022): »Digitising Biopiracy? The Global Governance of Plant Genetic Resources in the Age of Digital Sequencing Information«, in: Third World Quarterly 43 (8), pp. 1970–1987.
Nixon, Rob (2011): Slow Violence and the Environmentalism of the Poor, Cambridge, MA, London: Harvard University Press.
Nnadozie, Kent (2020): Global Goals and International Interdependence. How We All Depend on Each Other to Feed Ourselves, Longyearbyen, Svalbard.
NordGen (2008): Agreement Between (Depositor) and the Royal Norwegian Ministry of Agriculture and Food Concerning the Deposit of Seeds in the Svalbard Global Seed Vault, https://seedvault.nordgen.org/common/SGSV_Deposit_Agreement.pdf from July 2, 2025.
NordGen (2020): Svalbard Global Seed Vault Commences Seed Experiment that will last for 100 Years, https://www.nordgen.org/news/svalbard-global-seed-vault-commences-seed-experiment-that-will-last-for-100-years/ from May 27, 2024.
NordGen (n.d. a): Svalbard Global Seed Vault, https://www.nordgen.org/our-work/svalbard-global-seed-vault/ from July 18, 2025.
NordGen (n.d. b): Svalbard Global Seed Vault's Seed Portal, https://seedvault.nordgen.org/ from July 2, 2025.
Norgaard, Kari M. (2011): Living in Denial. Climate Change, Emotions, and Everyday Life, Cambridge, MA: MIT Press.

Norwegian Ministry of Agriculture and Food (n.d. a): Svalbard Global Seed Vault. History, https://www.seedvault.no/about/history/ from July 2, 2025.

Norwegian Ministry of Agriculture and Food (n.d. b): Svalbard Global Seed Vault. The Seeds, https://www.seedvault.no/about/the-seeds/ from July 2, 2025.

Norwegian Ministry of Climate and Environment (2001): Svalbard Environmental Protection Act (Amended by the Act of 20 April 2012 No. 20), https://www.regjeringen.no/en/dokumenter/svalbard-environmental-protection-act/id173945/ from July 10, 2025.

Norwegian Ministry of Foreign Affairs/Crop Trust/NordGen (n.d.): Svalbard Global Seed Vault. Virtual Tour, https://seedvaultvirtualtour.com/ from July 2, 2025.

Norwegian Polar Institute (n.d.): Climate Change in the Arctic, https://www.npolar.no/en/themes/climate-change-in-the-arctic/ from July 2, 2025.

Nuttall, Mark (Ed.) (2005): Encyclopedia of the Arctic, New York, London: Routledge.

Ojala, Maria/Cunsolo, Ashlee/Ogunbode, Charles A./Middleton, Jacqueline (2021): »Anxiety, Worry, and Grief in a Time of Environmental and Climate Crisis. A Narrative Review«, in: Annual Review of Environment and Resources 46 (1), pp. 35–58.

Øseth, Ellen (2011): Climate Change in the Norwegian Arctic. Consequences for Life in the North, Tromsø.

Paajanen, Annukka (2024): »Seeds in Translation. From Multispecies Environments into Frozen Assets«, in: Interconnections: Journal of Posthumanism 3 (1), pp. 20–38.

Pellegrini, Pablo A./Balatti, Galo E. (2016): »Noah's Arks in the XXI Century. A Typology of Seed Banks«, in: Biodiversity and Conservation 25 (13), pp. 2753–2769.

Peres, Sara (2016): »Saving the Gene Pool for the Future. Seed Banks as Archives«, in: Studies in History and Philosophy of Biological and Biomedical Sciences 55, pp. 96–104.

Peres, Sara (2019): »Seed Banking as Cryopower. A Cryopolitical Account of the Work of the International Board of Plant Genetic Resources, 1973–1984«, in: Culture, Agriculture, Food and Environment 41 (2), pp. 76–86.

Peters, Jochen (2005): »Svalbard Treaty«, in: Nuttall (Ed.), Encyclopedia of the Arctic, 1982–1983.

Philip, Kavita (2004): Civilizing Natures. Race, Resources, and Modernity in Colonial South India, New Brunswick: Rutgers University Press.

Pietikäinen, Sari (2024): Cold Rush. Critical Assemblage Analysis of a Heating Arctic, Cham: Palgrave Macmillan.

Pihkala, Panu (2024): »Ecological Sorrow. Types of Grief and Loss in Ecological Grief«, in: Sustainability 16 (849): pp. 1–44.

Pirli, Myrto/Schweitzer, Johannes/Ottemöller, Lars et al. (2010): »Preliminary Analysis of the 21 February 2008 Svalbard (Norway) Seismic Sequence«, in: Seismological Research Letters 81 (1), pp. 63–75.

Pistorius, Robin (1997): Scientists, Plants and Politics. A History of the Plant Genetic Resources Movement, Rome: IPGRI.

Plumwood, Val (1993): Feminism and the Mastery of Nature, London, New York: Routledge.

Plumwood, Val (2002): »Decolonizing Relationships with Nature«, in: William M. Adams/Martin Mulligan (Eds.), Decolonizing Nature. Strategies for Conservation in a Post-Colonial Era, Abingdon, New York: Earthscan Publications, pp. 77–110.

Potter, Melissa/Puckett, Maggie (2019): An Illuminated Feminist Seed Bank, http://www.seedsinservice.com/ from July 9, 2025.

Prescott, Tucker (2019): Offerings, https://vimeo.com/387107770 from June 2, 2024.

Puig de la Bellacasa, María (2017): Matters of Care. Speculative Ethics in More Than Human Worlds, Minneapolis: University of Minnesota Press.

Quijano, Aníbal (2007): »Coloniality and Modernity/Rationality«, in: Cultural Studies 21 (2–3), pp. 168–178.

Qvenild, Marte (2005): Securing Seeds in Permafrost. An Idea Whose Time Has Come. Master Thesis, Department of International Environment and Development Studies, Noragric, Norwegian University of Life Sciences (UMB). http://www.umb.no/statisk/noragric/publications/master/2006_marte_qvenild.pdf from June 10, 2024.

Qvenild, Marte (2008): »Svalbard Global Seed Vault. A ›Noah's Ark‹ for the World's Seeds«, in: Development in Practice 18 (1), pp. 110–116.

Qvenild, Marte (2012): »Svalbard Global Seed Vault«, in: Louis Kotzé/Stephen Morse (Eds.), Afro-Eurasia. Assessing Sustainability, Great Barrington: Berkshire Publishing Group, pp. 303–306.

Rabinow, Paul (1996): »Artificiality and Enlightenment. From Sociobiology to Biosociality«, in: Essays on the Anthropology of Reason, Princeton: Princeton University Press, pp. 91–111.

Radin, Joanna (2013): »Latent Life. Concepts and Practices of Human Tissue Preservation in the International Biological Program«, in: Social Studies of Science 43 (4), pp. 484–508.

Radin, Joanna (2017): Life on Ice. A History of New Uses for Cold Blood, Chicago: University of Chicago Press.

Radin, Joanna/Kowal, Emma (2017): »Introduction. The Politics of Low Temperature«, in: Cryopolitics, pp. 3–25.

Rae, Patricia (Ed.) (2007): Modernism and Mourning, Lewisburg: Bucknell University Press.

Ramirez, Marleni/Ortiz, Rodomiro/Taba, Suketoshi et al. (2012): »Demonstrating Interdependence on Plant Genetic Resources for Food and Agriculture«, in: Halewood/López Noriega/Louafi (Eds.), Crop Genetic Resources as a Global Commons, pp. 53–75.

Ramirez-Villegas, Julian/Jarvis, Andy/Fujisaka, Sam et al. (2012): »Crop and Forage Genetic Resources. International Interdependence in the Face of Climate Change«, in: Halewood/López Noriega/Louafi (Eds.), Crop Genetic Resources as a Global Commons, pp. 78–98.

Rappaport, Roy A. (1979): Ecology, Meaning, and Religion, Berkeley, CA: North Atlantic Books.

Rappaport, Roy A. (1999): Ritual and Religion in the Making of Humanity, Cambridge, UK: Cambridge University Press.

Ribeiro, Silvia/Shand, Hope (2008): »Seeding New Technologies to Fuel Old Injustices«, in: Development 51 (4), pp. 496–503.

Riesch, Hauke (2021): Apocalyptic Narratives. Science, Risk and Prophecy, Abingdon, New York: Routledge.

Roosth, Sophia (2016): »Virus, Coal, and Seed. Subcutaneous Life in the Polar North«, in: Los Angeles Review of Books from December 21, 2016, https://lareviewofbooks.org/article/virus-coal-seed-subcutaneous-life-polar-north/ from July 2, 2025.

Rose, Deborah B. (2011): Wild Dog Dreaming. Love and Extinction, Charlottesville: University of Virginia Press.

Rose, Nikolas S. (2007): Politics of Life Itself. Biomedicine, Power, and Subjectivity in the Twenty-First Century, Princeton: Princeton University Press.

Royal Ministry of Justice (1988): Treaty of 9 February 1920 Relating to Spitsbergen (Svalbard), Act of 17 July 1925 Relating to Svalbard, and The Mining Code for Spitsbergen (Svalbard), Oslo, Norway, https://app.uio.no/ub/ujur/oversatte-lover/data/lov-19250717-011-eng.pdf from July 9, 2025.

Ryan, John C. (2017): »Where Have All the Boronia Gone? A Posthumanist Model of Environmental Mourning«, in: Cunsolo/Landman (Eds.), Mourning Nature, pp. 117–143.
Sánchez Cedillo, Raúl (2023): Dieser Krieg endet nicht in der Ukraine. Argumente für einen konstituierenden Frieden [This War will not end in Ukraine. Arguments in Favour of Constituent Peace], Vienna: transversal texts.
Saraiva, Tiago (2013): »Breeding Europe. Crop Diversity, Gene Banks, and Commoners«, in: Cornelis Disco/Eda Kranakis (Eds.), Cosmopolitan Commons. Sharing Resources and Risks Across Borders, Cambridge, MA: MIT Press, pp. 185–211.
Sarmiento, Domingo ([1845] 1998): Civilization and Barbarism, New York: Penguin.
Sassen, Saskia (2007): A Sociology of Globalization, New York: Norton.
Schauer, Alexandra (2023): Mensch ohne Welt. Eine Soziologie spätmoderner Vergesellschaftung [Man Without World. A Sociology of Late Modern Sociation], Berlin: Suhrkamp.
Schiebinger, Londa (2004): Plants and Empire. Colonial Bioprospecting in the Atlantic World, Cambridge, MA: Harvard University Press.
Schiebinger, Londa (2005): »Prospecting for Drugs. European Naturalists in the West Indies«, in: Schiebinger/Swan (Eds.), Colonial Botany, pp. 119–133.
Schiebinger, Londa/Swan, Claudia (Eds.) (2005): Colonial Botany. Science, Commerce, and Politics in the Early Modern World, Philadelphia: University of Pennsylvania Press.
Schmink, Marianne/Wood, Charles H. (1992): Contested Frontiers in Amazonia, New York: Columbia University Press.
Schürkmann, Christiane (2021): »Facing a Toxic Object. Nuclear Waste Management and its Challenges for Nature-Culture-Relationships«, in: Nature and Culture 16 (1), pp. 65–82.
Scientific American (2008): »Preparing for Doomsday«, from May 1, 298 (5), p. 36.
Secretariat of the CBD ([1992] 2011): Convention on Biological Diversity. Text and Annexes. Opened for Signature on June 05, 1992, Montreal, Quebec, Canada, https://www.cbd.int/doc/legal/cbd-en.pdf from July 2, 2025.
Secretariat of the CBD (2005): Handbook of the Convention on Biological Diversity. Including its Cartagena Protocol on Biosafety, Montreal.
Seed Cultures Archive (2018): 2018 Deposit. Forgotten Stories of Frozen Seeds, https://www.seedcultures.com/svalbard-2019#/2018/ from July 2, 2025.

Seeds InService (2019): Seeds InService. Seed Cultures Archive, https://www.seedcultures.com/#/seeds-inservice/ from July 2, 2025.

Sempértegui, Andrea (2021): »Indigenous Women's Activism, Ecofeminism, and Extractivism. Partial Connections in the Ecuadorian Amazon«, in: Politics & Gender 17 (1), pp. 197–224.

Shiva, Vandana (1998): Biopiracy. The Plunder of Nature and Knowledge, Totnes: Green Books.

Shiva, Vandana/Jafri, Afsar H./Emani, Ashok/Pande, Manish (2002): Seeds of Suicide. The Ecological and Human Costs of Globalisation of Agriculture, New Delhi: Research Foundation for Science, Technology and Ecology.

Silberzahn, Léna (2024): »Care, Ecology, and the Crisis of Eco-Social Reproduction. Politicizing More-than-Human Care«, in: Hypatia 39 (4), pp. 711–731.

Smith, Joe/Clark, Nigel/Yusoff, Kathryn (2007): »Interdependence«, in: Geography Compass 1 (3), pp. 340–359.

Sokolickova, Zdenka (2022): »The Golden Opportunity? Migration to Svalbard from Thailand and the Philippines«, in: Nordic Journal of Migration Research 12 (3), pp. 293–309.

Spargo, R. C. (2004): The Ethics of Mourning. Grief and Responsibility in Elegiac Literature, Baltimore: Johns Hopkins University Press.

Star, Susan L. (Ed.) (1995): Ecologies of Knowledge. Work and Politics in Science and Technology, Albany: State University of New York Press.

Statistics Norway (2017): This is Svalbard 2016. What the Figures say, https://www.ssb.no/en/befolkning/artikler-og-publikasjoner/_attachment/294354?_ts=15a12de02c0 from July 2, 2025.

Steffen, Will/Broadgate, Wendy/Deutsch, Lisa et al. (2015a): »The Trajectory of the Anthropocene: The Great Acceleration«, in: The Anthropocene Review 2 (1), pp. 81–98.

Steffen, Will/Richardson, Katherine/Rockström, Johan et al. (2015b): »Planetary Boundaries. Guiding Human Development on a Changing Planet«, in: Science 347 (1259855), 1–10.

Stengers, Isabelle ([1997] 2010): Cosmopolitics I. 1. The Science Wars, 2. The Invention of Mechanics, 3. Thermodynamics, Minneapolis: University of Minnesota Press.

Stengers, Isabelle ([1997] 2011): Cosmopolitics II. 4. Quantum Mechanics, 5. In the Name of the Arrow of Time, 6. Life and Artifice, 7. The Curse of Tolerance, Minneapolis: University of Minnesota Press.

Strauss, Anselm L. (1978): »A Social Worlds Perspective«, in: Studies in Symbolic Interaction 1, pp. 119–128.

Strauss, Anselm L. (1987): Qualitative Analysis for Social Scientists, Cambridge, UK: Cambridge University Press.

Strauss, Anselm L./Corbin, Juliet M. (1990): Basics of Qualitative Research. Grounded Theory Procedures and Techniques, London: SAGE Publications.

Sturgeon, Noël (1997): Ecofeminist Natures. Race, Gender, Feminist Theory, and Political Action, New York: Routledge.

Styres, Sandra/Zinga, Dawn (2013): »The Community-First Land-Centred Theoretical Framework. Bringing a ›Good Mind‹ to Indigenous Education Research?«, in: Canadian Journal of Education 36 (2), pp. 284–313.

Subramaniam, Banu (2024): Botany of Empire. Plant Worlds and the Scientific Legacies of Colonialism, Seattle: University of Washington Press.

Sunder Rajan, Kaushik (2006): Biocapital. The Constitution of Postgenomic Life, Durham: Duke University Press.

Sustainable Development Goals Advocates (2020): Arctic Call to Action on Food Security and Climate Change. United Nations, Svalbard, Norway, https://www.unsdgadvocates.org/news/arctic-call-to-action-on-food-security-and-climate-change from July 2, 2025.

Svalbard Museum (n.d. a): Fur Hunting, https://svalbardmuseum.no/en/fur-hunting from July 2, 2025.

Svalbard Museum (n.d. b): Mining Towns. Longyearbyen, https://svalbardmuseum.no/en/longyearbyen from July 2, 2025.

Svalbard Museum (n.d. c): The Svalbard Treaty, https://svalbardmuseum.no/en/the-svalbard-treaty from July 2, 2025.

Svalbard Museum (n.d. d): Whaling, https://svalbardmuseum.no/en/whaling from July 2, 2025.

TallBear, Kim (2013): »Genomic Articulations of Indigeneity«. Special Issue: Indigenous Body Parts and Postcolonial Technoscience, in: Social Studies of Science 43 (4), pp. 509–533.

The Guardian (2015): »The Doomsday Vault. The Seeds that Could Save a Post-Apocalyptic World«, from May 20, written by Suzanne Goldenberg, https://www.theguardian.com/science/2015/may/20/the-doomsday-vault-seeds-save-post-apocalyptic-world from July 9, 2025.

The Migrant Ecologies Project (2019): Seeding Stories. A Guide to the Interior of a Salt Water Crocodile. Seed Cultures Archive, https://www.seedcultures.com/#/migrant/ from July 25, 2025.

The Verge (2017): »›Doomsday‹ Seed Vault Meant to Survive Global Disasters Breached by Climate Change«, from May 19, written by Alessandra Potenza, https://www.theverge.com/2017/5/19/15664298/svalbard-global-seed-vault-norway-doomsday-climate-change from July 9, 2025.

Thompson, Neil/Doka, Kenneth J. (2017): »Disenfranchised Grief«, in: Neil Thompson/Gerry R. Cox (Eds.), Handbook of the Sociology of Death, Grief, and Bereavement. A Guide to Theory and Practice, New York, London: Routledge, pp. 1–14.

Time (2018): »Norway's Underground Doomsday Seed Vault is Under Threat from Climate Change«, from February 27, written by Flora Carr, https://time.com/5177165/climate-change-threatens-norway-seed-vault/ from July 2, 2025.

Tsing, Anna L. (2003): »Natural Resources and Capitalist Frontiers«, in: Economic and Political Weekly 38 (48), pp. 5100–5106.

Tsing, Anna L. (2005): »How to Make Resources in Order to Destroy Them (and Then Save Them?) on the Salvage Frontier«, in: Daniel Rosenberg/Susan Harding (Eds.), Histories of the Future, Durham: Duke University Press, pp. 53–73.

Tsing, Anna L. (2015): The Mushroom at the End of the World. On the Possibility of Life in Capitalist Ruins, Princeton, Oxford: Princeton University Press.

Tsing, Anna L. (2019): »On Nonscalability. The Living World is not Amenable to Precision-Nested Scales«, in: Common Knowledge 25 (1–3), pp. 143–162.

Turner, Frederick J. ([1893] 1994): »The Significance of the Frontier in American History«, in: John M. Faragher (Ed.), Rereading Frederick Jackson Turner, New York: Henry Holt and Company, pp. 31–60.

Turner, Victor (1982): From Ritual to Theatre. The Human Seriousness of Play, New York: Performing Arts Journal Publications.

Turner, Victor (1995): The Ritual Process. Structure and Anti-Structure, New York: Aldine de Gruyter.

UN Human Rights Council (2018): United Nations Declaration on the Rights of Peasants and Other People Working in Rural Areas. UN Resolution 39/12, https://digitallibrary.un.org/record/1650694?v=pdf#files from July 2, 2025.

van Dooren, Thom (2009a): »Banking Seed. Use and Value in the Conservation of Agricultural Diversity«, in: Science as Culture 18 (4), pp. 373–395.

van Dooren, Thom (2009b): »Genetic Conservation in a Climate of Loss. Thinking with Val Plumwood«, in: Australian Humanities Review 46 (1), pp. 103–112.

van Dooren, Thom (2014): Flight Ways. Life and Loss at the Edge of Extinction, New York: Columbia University Press.

van Dooren, Thom (2019): The Wake of Crows. Living and Dying in Shared Worlds, New York: Columbia University Press.

van Dooren, Thom (2022): A World in a Shell. Snail Stories for a Time of Extinctions, Cambridge, MA: The MIT Press.

van Dooren, Thom/Kirksey, Eben/Münster, Ursula (2016): »Multispecies Studies. Cultivating Arts of Attentiveness«, in: Environmental Humanities 8 (1), pp. 1–23.

van Wyck, Peter (2005): Signs of Danger. Waste, Trauma and Nuclear Threat, Minneapolis: University of Minnesota Press.

Vavilov, Nikolaj I. ([1926] 1992a): »Centers of Origin of Cultivated Plants«, in: Origin and Geography of Cultivated Plants, pp. 22–135.

Vavilov, Nikolaj I. (Ed.) (1992b): Origin and Geography of Cultivated Plants, Cambridge, UK: Cambridge University Press.

Verran, Helen (2018): »The Politics of Working Cosmologies Together While Keeping Them Separate«, in: de la Cadena/Blaser (Eds.), A World of Many Worlds, pp. 112–130.

Visser, Bert (2012): »The Moving Scope of Annex 1. The list of Crops Covered under the Multilateral System«, in: Halewood/López Noriega/Louafi (Eds.), Crop Genetic Resources as a Global Commons, pp. 265–282.

Viveiros de Castro, Eduardo (1998): »Cosmological Deixis and Amerindian Perspectivism«, in: The Journal of the Royal Anthropological Institute 4 (3), pp. 469–488.

Viveiros de Castro, Eduardo (2014): Cannibal Metaphysics, Minneapolis: University of Minnesota Press.

Viveiros de Castro, Eduardo (2015): The Relative Native. Essays on Indigenous Conceptual Worlds, Chicago: HAU Books.

Viveiros de Castro, Eduardo/Danowski, Déborah (2018): »Humans and Terrans in the Gaia War«, in: de la Cadena/Blaser (Eds.), A World of Many Worlds, pp. 172–203.

von Verschuer, Franziska (2021): »Making Post/Anthropocentric Futures with Agrobiodiversity Conservation«, in: Nature and Culture 16 (1), pp. 47–64.

von Verschuer, Franziska/Hartmann, Viona/Barla, Josef (2023): »Um/Ordnungen des Lebens. Konturen einer Dekolonialen Analytik Ökologischer Krisen« [»Re/Ordering Life. Contours of a Decolonial Analytics of Ecological Crises«], in: Hoppe/Rüppel/von Verschuer/Voigt (Eds.), Leben Regieren [Governing Life], pp. 37–54.

Waehler, Turid A. (2022): »Hot Conflict in Cold Surroundings? Svalbard's Strategic Impact After Russia's Full-Scale Invasion of Ukraine«. Commentary, in: High North News from October 26, 2022, https://www.highnorthnews.com/en/hot-conflict-cold-surroundings-svalbards-strategic-impact-after-russias-full-scale-invasion-ukraine from July 2, 2025.

Wallace, Molly (2016): Risk Criticism. Precautionary Reading in an Age of Environmental Uncertainty, Ann Arbor: University of Michigan Press.

Wallerstein, Immanuel M. (1989): The Second Era of Great Expansion of the Capitalist World-Economy, 1730–1840s, New York: Academic Press.

Walsh, Catherine E. (2018): »Decoloniality in/as Praxis. Part I«, in: Mignolo/Walsh (Eds.), On Decoloniality, pp. 13–102.

Wehling, Peter (2011): »Vom Risikokalkül zur Governance des Nichtwissens. Öffentliche Wahrnehmung und Soziologische Deutung von Umweltgefährdungen« [»From Risk Calculation to the Governance of Nonknowledge. Public Perception and Sociological Interpretation of Environmental Hazards«], in: Matthias Groß (Ed.), Handbuch Umweltsoziologie [Handbook of Environmental Sociology], Wiesbaden: VS Verlag für Sozialwissenschaften, pp. 529–548.

Westengen, Ola T./Lusty, Charlotte/Yazbek, Mariana et al. (2020): »Safeguarding a Global Seed Heritage from Syria to Svalbard«, in: Nature Plants 6 (11), pp. 1311–1317.

Whatmore, Sarah (2002): Hybrid Geographies. Natures, Cultures, Spaces, London: SAGE.

Whitt, Laurie A. (1998): »Biocolonialism and the Commodification of Knowledge«, in: Science as Culture 7 (1), pp. 33–67.

Whyte, Kyle P. (2015a): »Indigeneity«, in: Joni Adamson/William E. Gleason/David N. Pellow (Eds.), Keywords for Environmental Studies, New York: New York University Press, pp. 143–146.

Whyte, Kyle P. (2015b): »Indigenous Food Systems, Environmental Justice, and Settler-Industrial States«, in: Mary C. Rawlinson/Caleb Ward (Eds.), Global Food, Global Justice. Essays on Eating under Globalization, Newcastle upon Tyne: Cambridge Scholars Publishing, pp. 143–156.

Whyte, Kyle P. (2017): »Is it Colonial Déjà Vu? Indigenous Peoples and Climate Injustice«, in: Joni Adamson/Michael Davis (Eds.), Humanities for the Environment. Integrating Knowledge, Forging new Constellations of Practice, London, New York: Routledge, pp. 88–105.

Wickson, Fern (2016): »Do We Care About Synbiodiversity? Questions Arising from an Investigation into Whether There are GM Crops in the Svalbard Global Seed Vault«, in: Journal of Agricultural and Environmental Ethics 29 (5), pp. 787–811.

Wired (2008): »The (End of the) World Seed Vault Opens«, from February 29, written by Alexis Madrigal, https://www.wired.com/2008/02/the-end-of-the/ from July 9, 2025.

Wired (2017): »The Arctic Doomsday Seed Vault Flooded. Thanks, Global Warming«, from May 19, written by Damien Carrington, https://www.wired.com/2017/05/arctic-doomsday-seed-vault-flooded-thanks-global-warming/ from July 2, 2025.

Wolff, Leon (2020): »The Past Shall Not Begin. Frozen Seeds, Extended Presents and the Politics of Reversibility«, in: Security Dialogue 7 (1), 1–17.

Wolff, Leon (2024): »Suspending Life, Controlling Change. Cryotechnology, Genetic Identity, and Ecological Separation«, in: BioSocieties 19 (4), pp. 574–594.

World Commission on Environment and Development (1987): Our Common Future, Oxford: Oxford University Press.

Wynter, Sylvia (2003): »Unsettling the Coloniality of Being/Power/Truth/Freedom. Towards the Human, After Man, Its Overrepresentation – An Argument«, in: The New Centennial Review 3 (3), pp. 257–337.

Yusoff, Kathryn (2018): A Billion Black Anthropocenes or None, Minneapolis: University of Minnesota Press.

Zalasiewicz, Jan/Williams, Mark/Smith, Alan et al. (2008): »Are We Now Living in the Anthropocene?«, in: GSA Today 18 (2), pp. 4–8.

Žižek, Slavoj (2022): »From Catastrophe to Apocalypse... and Back«. Apocalyptica 1 (1), pp. 36–53.

Zylinska, Joanna (2018): The End of Man. A Feminist Counterapocalypse, Minneapolis: University of Minnesota Press.